Primitive Accumulation and Community

Ryoichi Yamazaki

Tsukubashobo

Primitive Accumulation and Community
Ryoichi Yamazaki

First Publishing 2024 by Tsukuba-shobo
Publisher : Haruhiko Tsurumi
Tsukuba-shobo Publishing Company
2-16-5 Kagurazaka Shinjuku-ku Tokyo Japan 162-0825
Phone : 03-3267-8599
ISBN978-4-8119-0688-1

Preface

This book deals with theoretical issues in historical economics. The book begins by reexamining two concepts that have long been used in the history of academic theory: primitive accumulation (Chapters 1-3) and community (Chapters 4 and 5), and then attempts to offer my own original interpretations of them. In addition, the book proposes new analytical frameworks, including the typology of process of primitive accumulation (Chapter 6), the reproduction schema incorporating non-capitalist surroundings (Chapter 7), and the historical view from primitive accumulation (Final chapter). These proposals have been made possible by the book's examination of the two concepts mentioned above, as well as by empirical research conducted over many years in Japan, Vietnam, Mali and France. This book is also a theoretical summary of such empirical research. If I have succeeded in saying something different from previous scholars in this book, it is solely due to the method I have used to construct theory from reality. Needless to say, the discussions of previous scholars served as a guiding light for me in this process. In particular, if I had not come across Karl Marx's "Pre-Capitalist Economic Formations" when I was young, I probably would not have traveled all the way to the banks of the Niger River in West Africa to study pre-capitalist communities and conduct systematic surveys of settlements there.

In addition, I was greatly inspired by the discussions I had with the collegians. In particular, one of the main points of discussion with my dear friend, Professor Masahiro Yamamoto, was the interpretation of the concept of primitive accumulation. In addition, my academic senior, Professor Norio Tsuge, read through the entire volume and provided me with many useful comments, which helped me to improve the descriptions in many parts of the text. In some places, I have adopted comments from Professor Tsuge verbatim into the text. Such points are noted in the text. Professor Tsuge has also provided me with valuable suggestions regarding the direction of research into the issues that will be necessary in further developing the reproduction schema incorporating non-capitalist surroundings and the historical view from primitive accumulation, both of which are highly tentative in nature.

The Japanese book on which this book is based was published as Part 1 of Volume 5 of my Collected Works. Fortunately, this Japanese book won the Academic Award of the Agricultural Economics Society of Japan in 2023. I would like to thank those who recommended this book and those who reviewed it.

Finally, I would like to take this opportunity to express my gratitude to President Haruhiko Tsurumi of Tsukuba Shobo, who has always supported me in the publication of my Collected Works from 2020 to 2022 and this book, as well as to Professor Sachiho Arai of the Agricultural Economics Department at Tokyo University of Agriculture and Technology, who came up with the original project for this collected wowks, and to the other members of the publication committee.

Sapporo, Japan
Ryoichi Yamazaki
August 2024

Contents

Introduction : Issues and Perspective

As a student of agricultural economics, I have long been engaged in survey research on realities of regions. The journey up to this point can be roughly divided into the following three periods. The first is the period from the late 1980s to the early 1990s, targeting research sites in Japan. The researches that resulted in the papers were conducted in the following target areas. (1) Kitaura Village,[1] Namekata District, Ibaraki Prefecture (1987), (2) Azuma Village,[2] Inashiki District, Ibaraki Prefecture (1988), (3) Kameyama City, Mie Prefecture (1990), (4) Miyada Village, Kamiina District, Nagano Prefecture (1993). However, the period after Yamazaki (1996) was completed based on these research results was the second period of my research. There, I turned my focus to foreign countries and traveled to France, the Mekong Delta in Vietnam, and the Inner Niger Delta in Mali. During this period, I stayed at the Toulouse branch of INRA (French National Institute for Agricultural Research: August 1994-October 1995), Can Tho University in Vietnam (September 1996-August 1997), and CIRAD (French Agricultural Research Centre for International Development: April 2003-March 2004) and spent a long time abroad. The research results of this period have already been published as Yamazaki (2007). The third period in recent years is after I moved to Tokyo University of Agriculture and Technology in 2009. Together with the staff of the Agricultural Economics Laboratory and young students, I returned to Japan and have been working on a follow-up survey in Kamiina District, Nagano Prefecture (since 2009) and a survey in Higashiomi City,

1 Currently part of Namekata City.
2 Now part of Inashiki City.

Shiga Prefecture (2010).[3] On the other hand, research in southern Vietnam (since 1999) is also continuing.

I have been conducting surveys and research closely related to the region, even with the transition of the target area, such as domestic → overseas → both. Through this process, I tried to think about the problem framework of "the relationship between capital and agriculture," and analyzed the structural dynamics of agriculture from this perspective. However, this framework of problems is by no means unique to me. On the contrary, it is an orthodox one that has long been important in Japanese agricultural economics. The guideline there was to elucidate the mechanism by which "forces outside agriculture are transformed into constituent parts as internal factors within agriculture" (Yamada 1962). However, it seems that this problem framework has not been taken up as often as it used to be in recent years in Japanese agricultural economics. On the other hand, tensions and conflicts between huge capital and peasant agriculture[4] seem to be increasing in recent years. One manifestation of this is the widening gap between the economic and political power of peasant agriculture and huge capital, which is progressing on a global scale. One manifestation of this is the widening gap between the economic and political power of peasant agriculture and huge capital, which is progressing on a global scale. Agriculture is responsible for food production and is the most fundamental and important industry for human existence as a living organism. In spite of this, today the farmers, who are the main agents of agriculture, appear to be groaning under the pressure of huge capital while being forced into unfavorable terms of trade between farmers and industries. In the face of such a situation, am I the only one who wants to put the problem framework

---

3 I have also conducted surveys in Inashiki City, Yokote City in Akita Prefecture, Goshogawara City in Aomori Prefecture, and Okinawa Prefecture.

4 "Peasant agriculture" is farming as an industry carried out by small farmers.

of "the relationship between capital and agriculture" on the table?

In modern capitalist society, it is undeniable that peasant agriculture exists under the great influence and pressure of the huge capital that is dominant in this society. Approaching from the problem framework of "relationship between capital and agriculture" by previous studies has aimed at clarifying "how" of this decisive influence. Furthermore, in the extension of the elucidation of this decisive influence, there must have been an awareness of the necessity of repelling the pressure from huge capital and sticking to the subjective logic of peasant agriculture. This is because the logic of huge capital distorts the logic of the development of agricultural productivity, which is compatible with the improvement of the livelihood of farmers and the maintenance and enhancement of soil fertility that should be undertaken by farmers (Shiina 1976). Therefore, the analysis of the process of the former's grasp of the latter was at the same time an attempt to see the path through which the subjectivity of peasant agriculture, which was opposed to the logic of huge capital, would be achieved. Therefore, the analysis of the process of taking control of farmers by capital was also the prospect of a path to implement the subjectivity of peasant agriculture, which is opposed to the logic of huge capital. Conventionally, research based on the problem framework of "the relationship between capital and agriculture" has been conducted under this perspective. By the way, such research cannot be fully accomplished only by collecting data through fact-finding surveys and searching for statistical data, and then organizing and aggregating those data. In addition, it demands the theoretical methodology necessary to analyze the data. Agricultural market studies is one of the fields of agricultural economics. This was originally an approach to agricultural problems from the problem framework described here (Mishima 2005). Looking at five markets (labor market, land market, farmers' purchases market, financial market, agricultural products market) as places where huge capital and peasant agriculture establish an economic relationship, the relationship between the former and the latter, and the logic of the latter's opposition to the former, have been studied for many years. However, what I have gradually become aware of in pursuing agricultural market theory is that this theory must be premised on a systematic understanding of the capitalist society. In other words, on what logic does the capitalist society operate, and why should the huge capital, which is in a dominant position in this society, have a relationship with peasant agriculture through the five markets? Not only the "how" of the impact of huge capital on peasant agriculture, but also the "why" must be raised as a problem at the same time. This problem naturally demands a systematic understanding of capitalist society. It seems to me that agricultural market studies do not give a sufficient answer to such a seemingly "naive" question, since the discussion begins with the assumption of five markets. In this sense, agricultural market studies may have been a system that emphasized understanding of phenomena. Alternatively, when there is strong sympathy for peasant agriculture, giant capital has been regarded as a villain who uses peasant agriculture for the pursuit of profit. Therefore, although phenomenology and this kind of theory of the promotion of good and punishment appear to be very different at first glance, they are actually very similar in that they neglect deep insight into the essence of the capitalist society. These seemed to me to be two phenomenal forms of the same substance. I have been carrying out "creeping on the ground" research on the other side of the earth, while at the same time continuing my theoretical studies from the viewpoint of "why" do huge capital and peasant agriculture come into contact with each other. By the way, when considering the question of "why" come into contact between huge capital and peasant agriculture, the direction in which the argument develops varies greatly depending on which side of the two the active subjectivity is placed.

However, when making this choice in research on the capitalist society, it is inevitable to see the active subject on the side of the huge capital that objectively leads this society. Therefore, it is necessary first to elucidate the logic of huge capital, and to find out where the impetus to combine with peasant agriculture lies within it.

At such times, we are tempted to return to "Capital," which provided a systematic understanding of the capitalist society. However, "Capital," a classic from the middle to the latter half of the 19th century, is an elucidation of the internal logic of the capitalist society in the so-called liberal stage, before the establishment of huge capital. What should be read from that book is the more general logic of the relationship between the capitalist sector and peasant agriculture. By the way, when we read "Capital" from this point of view, we immediately notice that the logic of the relationship between the capitalist sector and peasant agriculture is not fully developed. It is true that in Volume 3, Part 6 of "Capital" in the current edition, there is a so-called rent theory titled "Conversion of excess profits into rent." As is well known, there Marx directly critically succeeds Ricardo's theory of rent and discusses the problem of economic realization of land ownership as rent, which arises when capital captures agriculture. However, as Marx himself said, the subject of land rent theory is basically the logic of the relationship between capitalist agriculture and land ownership, not the logic of non-capitalist peasant agriculture. Nor is it the logic of the relationship between capital and peasant farming. However, Chapter 47, Section 5, titled "Share farming and peasant land ownership," is an exceptional example of Part 6 and develops the logic of peasant agriculture in a capitalist society. However again, this is not a discussion of the relationship between the two, "why" the capitalist sector comes into contact with peasant agriculture. Instead, it is the logic of agricultural product price formation and its rent formation inherent in peasant management. As is well known, not only the theory of land rent, but the basic logic of "Capital" as a whole is discussed on the assumption that society as a whole is covered by capitalist relations. The logic of the non-capitalist mode of production is therefore outside the subject of this classic. Therefore, trying to find out in "Capital" the logic of the relationship between the capitalist sector and peasant agriculture will not escape the accusation that it is an empty request. If we find it difficult to find clues in "Capital," from what angle should we think about the relationship between capital and peasant agriculture? Moreover, I would like to explain this problem from the most fundamental point. In this case, even though "Capital" itself does not take this relationship as its subject, the clue is still in this great book.

By the way, a capitalist society is a society based on the capital-labor relationship, and this point is a remarkable feature of the society when compared with societies based on other modes of production. Therefore, in considering why the capitalist sector comes into contact with peasant agriculture, we should consider what kind of role peasant agriculture plays in the formation and development of the capital-labor relationship. By the way, farmers are transformed into workers by being employed by capital through the labor market. Therefore, in the relationship between capital and peasant agriculture, among the above five markets, the relationship through the labor market seems to be of primary importance. Let's call this emphasis the labor market foundation theory in agricultural market studies.

By the way, I believe that the conventional general understanding of the relationship between the capitalist sector and peasant agriculture through the labor market, based on the classic, is as follows. However, there has been a separate discussion on the period of primitive accumulation[5] and the period

---

5 Sometimes called original accumulation.

after the establishment of the capitalist society. Among them, during the primitive accumulation period, labor-power is released from peasant agriculture through disintegration through mutual competition among peasants or through land expropriation from peasants by the ruling class using various forms of violence. Labor-power is then absorbed into the capitalist sector. At the same time, in this process, the community, which was the basic unit of pre-capitalist society, is shattered, and this situation is an important element of primitive accumulation. On the other hand, after the establishment of the capitalist society, the increase of the organic composition of capital and the creation of a relative surplus-population through the economic cycle created a labor force procurement mechanism peculiar to this society as the law of capitalist population. The role of peasant agriculture is no longer a one-sided supply source of labor, but also a pool of surplus-population during depression. The consideration of the two concepts (primitive accumulation and community) in this book is premised on the recognition of the above understanding of the relationship between the capitalist sector and peasant agriculture over the labor market. However, I tried to understand this basic recognition in a more concrete manner while giving answers to some of my questions that still remain. As a result, my understanding of these concepts as written in the chapters is in some ways distant from the public understanding.

In this book, after conducting a study aimed at elaborating the above concepts, I recognize the diversity of the primitive accumulation process among countries, and categorize this process among countries. This work is premised on the prediction that the national characteristics of the control of the peasants through the labor market by capital are acutely manifested in the primitive accumulation whose essential content is the separation of peasants and land. Since the process of primitive accumulation is located at the starting point of capitalist society, it goes without saying that its nature determines the initial characteristics of each society. However, even after the establishment of capitalist societies, it seems possible that the nature of the primitive accumulation process is a circumstance that determines the overall pattern of each country's capitalist society at its root. The characteristics of each country's primitive accumulation process, which is the process of transition from farmers to workers, are strongly reflected in the characteristics of the development process after the establishment of capitalist society, in which the relationship between capital and wage labor is the most basic social relationship. Today, as globalization progresses, the characteristics of each country's capitalist society seem to be unified, but I will try to identify the origin of the historical individuality of each country's society, which still continues to assert itself.

Furthermore, this typification work is also to develop the following description of primitive accumulation in the current "Capital."

"The expropriation of the agricultural producer, of the peasant, from the soil, is the basis of the whole process. The history of this expropriation in different countries, assumes different aspects, and runs through its various phases in different orders of succession, and at different periods. In England alone, which we take as our example, has it the classic form." (Marx 1867)

The first edition of the French version of "Capital," which is said to be "the last 'Capital'" because Marx himself modified it at the end of his life, was published in separate volumes in Paris in 1872-75. In the French version of "Capital," the following sentence is added immediately after the sentences corresponding to the above sentences in the current edition. Marx himself makes it clear that the description of the primitive accumulation in this classic, taking England as an example, can be applied, albeit with variations, among the countries of Western Europe.

"But all the other countries of Western Europe go through the same movement, although according to the environment it changes local color, or tightens in a narrower circle, or presents a character less strongly pronounced, or follows a different order of succession." (Marx 1872-1875)

This territorial limitation was later also quoted and emphasized by Marx himself in a famous letter from Marx to Zasulich (Marx 1881). This series of descriptions by Marx indicates that the primitive accumulation is diverse among countries and that it is possible to categorize it, and gives suggestions for future research.

In this book, based on the conceptual arrangement and categorization described above, I propose two new analytical frameworks: Schema of extended reproduction incorporating "non-capitalist surroundings," and the historical view from primitive accumulation. In the remainder of this chapter, I will outline the contents of each chapter in this book, focusing on my awareness of the problem.

In Chapter 1, "theoretical reexamination of primitive accumulation concept; focusing on the issue of creating a *Vogelfrei* labor force," the concept of primitive accumulation is reexamined based on the classical texts of economics. Primitive accumulation is, of course, one of the key concepts in this book, but, unfortunately, this concept seems to have been vague so far. Therefore, my motivation for writing this chapter is to understand this concept as clearly as possible. In this chapter, I will show the perspective of understanding the concept of primitive accumulation as the process of creating the preconditions for the functioning of the law of population in the capitalism. This is different from the conventional economic historical understanding of the concept of primitive accumulation as the historical formation process of the capitalist society, and rather emphasizes its theoretical significance in the system of economics. By the way, primitive accumulation is a concept that indicates the social movement of labor-power from the non-capitalist sector (typically farmer's agriculture) to the capitalist sector. This process itself consists of two elements. One is the process in which the pre-capitalist community that protected the peasants was destroyed by the penetration of the commodity economy and the violence of the ruling classes. The other is the process of differentiation and decomposition of the farmer class. In other words, there is a community at the starting point of this process, but what is a community? Conventionally, on the one hand, there has been the view that the community is regarded as a regulation that constrains the freedom of individual productive activities. On the other hand, there are also discussions focusing on regional resource management in the community, such as the discussion on the water user community, which has become popular in recent years in Japan. This seems to be an attempt to view the regulation by the community in a rather positive way from the perspective of the formation of agricultural productivity. Accordingly, does a community have regulation as its essential content? By the way, all of these arguments seem to be characterized by the fact that they emphasize the negative and positive aspects of the Japanese community while using it as a material. However, is the Japanese community really a typical community? Chapter 4, "community as a social system," draws on the one hand from some classical literature. On the other hand, I use the communities found in the Inner Niger Delta region, which is one of my research sites, as material for theorization. This chapter attempts to present an image of the community as a mutual aid economic system, which aims at protecting and relieving the members, and is therefore different from the community regulation as the main motive.

By the way, if we clearly understand the concept of primitive accumulation, we will be able to determine the timing of its "beginning" and "end." However, clarifying the concept of primitive accumulation does not only have an effective operational significance in analyzing the actual conditions of capitalist societies in each country. Moreover, this work reveals that the process of primitive accumulation was not merely the historical process of separation between the direct producer and the means of production, but has a deeper significance in human history. In other words, under primitive accumulation, the "symbiotic principle" of the pre-capitalist community was finally buried, competition among workers became the norm, and labor became decisively subordinate to capital. This is described in Chapter 2, "starting point and ending point of primitive accumulation," although the order of introduction is mixed up.

My understanding of the concept of primitive accumulation, which emphasizes the loss of land by direct producers, is criticized for neglecting the "accumulation of monetary wealth that can be converted into capital." Chapter 3, "on the controversy over the concept of primitive accumulation," shows that such criticism may actually be based on the incorrect description in Rosenberg's "Commentary on Capital." In Chapter 24 of "so-called primitive accumulation," Volume 1 of "Capital," I think that (1) the concept of primitive accumulation itself is distinguished from (2) "the elements of primitive accumulation." The "accumulation of monetary wealth" (but not the initial "accumulation of monetary wealth" converted into capital, but its additional accumulation that mobilized the mercantilist policy system) is one element of (2).

By the way, in a society in the period of primitive accumulation, a considerable part of the cost of reproducing labor-power can be passed on to the non-capitalist sector centering on peasant agriculture. As a result, the level of wages paid by capital when hiring workers can be kept lower than in the case where the labor force reproduction cost is not borne by the non-capitalist sector at all. In Chapter 5, "community and national differences of wages," this problem is theoretically examined as a low-wage problem in developing countries. What I particularly wanted to emphasize was the point that has tended to be overlooked in conventional discussions on low wages in developing countries, namely the effect of the existence of the community on wage levels.

In short, Chapters 1 to 5 focus on the community, but Chapter 6 is a geographical relativization of these discussions. First, in Section 2, based on the consideration of community in Chapter 4, I will reconsider and expand on the concept of primitive accumulation that I set forth in Chapters 1 to 3. In doing so, I consider that the spatial and historical overview of primitive accumulation consists of two mutually complementary processes; the "core" process in developed countries and the "peripheral" process in developing countries (Fujise 1985). Section 3 goes beyond that and typifies the primitive accumulation process of the "periphery" in modern developing countries. One is the Sub-Saharan African type, which requires a strong awareness of the social existence of a community, and the other is the Southeast Asian type, which does not require awareness of the social existence of a community.

At the beginning of the twentieth century, Luxemburg asserted that the capitalist society is dependent on the supply of labor from non-capitalist spheres. On the other hand, Luxemburg, while critically examining Marx's theory of reproduction scheme, also develops the famous theory of the unrealization of surplus-value. Somehow, however, she does not treat the non-capitalist realm as a source of labor-power in her theory of reproduction scheme. In the first place, not only in Luxemburg, but also in conventional reproduction scheme theory, the problem of the labor market has not been dealt with. Therefore, in Chapter 7, I remade the reproduction scheme by incorporating labor supply from the non-capitalist realm.

In England in the first half of the 19th century, (a) primitive accumulation, (b) the modernization of land ownership, and (c) the establishment of the national economy (industrial structure) synchronously integrated, bringing about the establishment of a capitalist society. In Japan, which is an example of a latecomer capitalist country, these three elements are separated in time, and among them, the accomplishment of primitive accumulation is entangled in the Heisei recession (1990s). If the "central" part of the primitive accumulation process is rich in nuances for each country, then the "peripheral" part of the process is also diverse. Based on the relative weight of the community in society, this book presents the Southeast Asian type and the Sub-Saharan African type. If I was able to develop such a typology of primitive accumulation, it would be because I clearly confirmed the contents of the two concepts of primitive accumulation and community. After organizing the history of capitalist societies around the world based on the concept of primitive accumulation, the final chapter expands on the understanding of the development stages of world capitalism.

Chapter 1 : Theoretical Reexamination of Primitive Accumulation Concept; Focusing on the Issue of Creating a *Vogelfrei* Labor Force

1. Theme

Primitive accumulation is originally a concept in economic history. At the same time, however, it is highly suggestive for considering the current situation in developing countries, and is therefore of modern significance. In short, this concept is quite a convenient one, but it is fraught with ambiguity. Therefore, the problem consciousness of this chapter is to reflect on this concept and to understand it as clearly as possible. Where, then, is the ambiguity? These are the following points. In general, the primitive accumulation as a historical process is regarded as the starting point of capitalist society and the process of transition from feudalism to capitalism. It is said that a capitalist society is established through primitive accumulation. For the time being, the understanding of primitive accumulation may be as described above. However, in this seemingly simple description, there is actually a content that is difficult to understand. And the difficulty is related to what it means to "establishment of a capitalist society."

What is the "establishment of a capitalist society?" From the point of view of reproduction theory, there is a position that defines it as "the establishment of a full-fledged reproduction of social total capital consisting of the production sector of production means and the production sector of consumer goods" (Yamada 1934). Alternatively, there is a position that "establishment of the capitalist law of population through the increase of the organic composition of capital" (Uno 1953), while viewing the fundamental contradiction of the capitalist society in terms of the commodification of labor-power. Both of these were realized in the early nineteenth century in Britain, the mother country of the capitalist system. Moreover, in England, it was also the time when the social transition from feudalism to capitalism was completed. Therefore, in England, it may be possible to assert the integral character of these three occasions on the basis of their concurrent nature.

However, in latecomer capitalist countries other than Great Britain, these three motives decompose into separate periods. For example, in Japan, the "settlement of the reproductive trajectory of total social capital" can be considered to have occurred in the 1910s when the heavy industry sector was established. There is also a well-known argument that the transition of society from feudalism to capitalism can be attributed to the post-World War II reforms, depending on the evaluation of the landowner-tenant relationship. In addition, the "establishment of the capitalist law of population through the increase of the organic composition of capital," I believe, occurred during the recession in the 1990s. (Yamazaki 2014a) As can be seen in this example, in the case of latecomer capitalist countries, the timing and end point of primitive accumulation differ considerably depending on which of the "establishment of a capitalist society" is regarded as the essential moment.

In other words, the question of what is the primitive accumulation leads to the question of what is the "establishment of a capitalist society." And this question comes down to another question, which of the above three moments should be considered as the standard. In order to clarify these questions, methodologically, it is necessary to first examine the descriptions in "Capital." Needless to say, the concept of primitive accumulation itself derives from "Capital."

What has been said about the primitive accumulation theory in Chapter 24 of the current Vol. 1, "So-called primitive accumulation?" There, the critical significance of Marx's primitive accumulation theory against the idyllic character of Adam Smith's so-called "previous accumulation," that is, capital ownership based on self-labor, has been emphasized (Marx 1867). In fact, Marx sees primitive accumulation as "the historical process of divorcing the producer from the means of production," and emphasizes the significance of the various forms of violence organized in this process. However, I believe that the primitive accumulation theory has a theoretical significance in terms of the "capital theory" system, which is not limited to such criticism of economics. What then is its significance? Let's look at Uno (1962) as a clue for thinking about this problem. According to him, based on the basic idea that "Capital" should be refined as the fundamental theory of economics, the primitive accumulation theory (Chapter 24) as a historical description is a supplement to the theory of capital accumulation in Chapters 21-23. In fact, I also agree with the point that the theory of primitive accumulation is a supplement to the theory of capital accumulation. However, I do not agree with the idea that the primitive accumulation theory is a supplement to the fundamental theory of economics because it is a historical description and is an extraneous part of the "Capital" system. Instead, as I mentioned earlier, I believe that the primitive accumulation theory has a positive significance in terms of the "Capital" system. What, then, is its significance? In order to give an answer to this question, this chapter begins by grasping the logical structure of the theory of capital accumulation in "Capital." This is because I think that clarifying where in this logical structure the concept of primitive accumulation should enter will clarify the theoretical significance of this concept. Through the clarification of the theoretical significance of the concept of primitive accumulation achieved through such work, I will also provide an answer to the question posed at the beginning of this chapter, "what is primitive accumulation?"

2. Issues in the Theory of Capital Accumulation

However, I will not put the entire theory of capital accumulation in "Capital" into consideration here. Here, I will focus on the description in Chapter 23 of Vol.1. For I consider the derivation of the capitalist law of population to be the subject of the theory of capital accumulation in the first volume. Let us now turn our attention to how the logic of Chapter 23 proceeds. In this chapter, Section 1 discusses "the increased demand for labor-power that accompanies accumulation, the composition of capital remaining the same," and Section 2 proceeds to the theory of the creation of excess population accompanying the "relative diminution of the variable part of capital." What I would like to raise as a question here is why the logic is carried out in this way. In other words, why does Chapter 23 discuss "accumulation in the case of unchanged capital composition" first, and then introduce the motive of "advancement of capital composition?" From the following description in "Capital," we can see that this way of carrying out logic is actually a reflection of the actual historical process. At least, that's what Marx thought.

"This peculiar course of modern industry, which occurs in no earlier period of human history, was also impossible in the childhood of capitalist production. The composition of capital changed but very slowly. With its accumulation, therefore, there kept pace, on the whole, a corresponding growth in the demand for labor." (Marx 1867)

Why, then, did such a development from "unchangeable capital composition" to "raised capital composition" actually occur? Or what are the historical conditions for this development to happen? Marx himself seems to have said nothing about this question. In other words, neither the historical conditions for this development nor the logical necessity are shown. On the other hand, Marx says that this situation of "childhood of capitalist production" was reflected in the description of "The Wealth of Nations" as follows.

"According to the economists themselves, it is neither the actual extent of social wealth, nor the magnitude of the capital already functioning, that lead to a rise of wages, but only the constant growth of accumulation and the degree of rapidity of that growth. (Adam Smith, Book I., Chapter 8.) So far, we have only considered one special phase of this process, that in which the increase of capital occurs along with a constant technical composition of capital." (Marx 1867)

Immediately after these sentences, however, Marx argues that Smith also focused on the development of labor productivity. Then, why did Smith not include this element in his theory of capital accumulation, even though he had the improvement of labor productivity in his view, and why did he place "the constant technical composition of capital" as a premise? Despite this, why was Marx able to place raise of technical composition of capital as a key concept in his theory of capital accumulation? This development of logic from Smith to Marx (development A) overlaps with Marx's own logical development (development B) from "constant capital composition" to "raised capital composition." Therefore, in the following, let us consider the question of how inevitable the logical development from Smith to Marx was. In other words, I would like to clarify the reason why "development B" was inevitable through considering the necessity of "development A." I think that Marx's internal logical development reflects the theory-historical logical development from Smith to Marx. In this way, the examination of Marx's theory of capital accumulation shifts to a comparative study of "Capital" and "The Wealth of Nations." Regarding the relationship between the two classics, Marx's criticism of Smith at the value theory level, which will be touched on later, has been emphasized (Uchida 1961). In this chapter, unlike that, we will examine the subject of the theory of primitive accumulation and capital accumulation.

By the way, I think that the progress in the theory of capital accumulation from Smith to Marx reflects the difference in the image of the worker that exists between them. Furthermore, I think that this difference in worker's image appears sharply when the two economists deal with the problem of inter-regional disparities in wages. Therefore, in the following, I would like to explore the difference in worker image between the two classics while examining the problem of regional wage disparities. As we will see later, it will become clear that Marx's "discovery" of (1) the primitive accumulation and (2) the historicity of the capitalist society lies behind this difference. I believe that the significance of the primitive accumulation theory as a criticism of economics must be understood in this dimension.

3. Regional Disparities of Wages in "The Wealth of Nations"

A comprehensive description of wages in the Wealth of Nations can be found in Chapters 8 through 10 of Volume 1. In this section, we will examine the descriptions while focusing on the problem of regional wage disparities.

First, let us quote the following description in Book 1, Chapter 8, "the wages of labor."

"But the wages of labor in a great town and its neighborhood are frequently a fourth or a fifth part, twenty or five-and-twenty per cent higher than at a few miles distance. Eighteen pence a day may be reckoned the common price of labor in London and its neighborhood. At a few miles distance it falls to fourteen and fifteen pence. Ten pence may be reckoned its price in Edinburgh and its neighborhood. At a few miles' distance it falls to eight pence, the usual price of common labor through the greater part of the low country of Scotland, where it varies a good deal less than in England. Such a difference of prices, which it seems is not always sufficient to transport a man from one parish to another, would necessarily occasion so great a transportation of the most

Table 1-1 Regional wage gap in unskilled labor (Smith)

England	London (city)	⇔	(farm village)
	a) 18p/day	(several miles)	b) 14-15p/day
		a-b = 3-4p/day	
Scotland	Edinburgh (city)	⇔	(farm village)
	c) 10p/day	(several miles)	d) 8p/day
		c-d = 2p/day	

① Wage gap between urban and rural areas : a > b , c > d
② Wage gap between England and Scotland : a > c, b > d
③ Difference between England and Scotland in wage inequality between urban and rural areas : (a-b) > (c-d)
(Source) Adapted from Chapter 8 of Smith (1776) .

bulkiest commodities, not only from one parish to another, but from one end of the kingdom, almost from one end of the world to the other, as would soon reduce them more nearly to a level. After all that has been said of the levity and inconstancy of human nature, it appears evidently from experience that a man is of all sorts of luggage the most difficult to be transported." (Smith 1776)

The main point raised here is (1) the "urban>rural" wage gaps that permeate both England and Scotland, with wages falling by 20-25% a few miles away from urban suburbs. (**Table 1-1**) However, along with this main point of contention, in fact, the following points of contention are also pointed out in this article. (2) Wage disparities as ⟨England>Scotland⟩ that permeate both urban and rural areas.[6] (3) the ⟨England>Scotland⟩ relationship seen in the urban-rural wage gap itself. The subject here is "ordinary labor," that is, unskilled labor. As for the factors that explain point (1), in this quote, "difficulty in transporting people" is pointed out as a general background. However, we must go further and ask what specific factors bring about this "immobility."

However, before proceeding to examine this problem, it is necessary to first consider Smith's discussion of wage level determination, that is, wage

6 Hunt (1986) on regional wage disparities between England and Scotland and their convergence process after the Industrial Revolution.

determination theory. This is because the problem of wage disparity between regions is a problem of how wage determination differs among regions, and therefore it can be understood as an application problem of wage determination theory. The theory of wage determination in "The Wealth of Nations" is largely discussed in the context of the theory of value. While his theory of value included the decomposition theory of value that Ricardo later inherited, it also sometimes stood in the position of the theory of value construction, and in short, was inconsistent. Marx pointed out this blur a long time ago. In any case, the theory of wage determination in "The Wealth of Nations" is discussed in the context of the theory of value construction. In other words, it is the theory of determining the average rate of wages, which is one of the factors that constitute the value of commodities. Its contents are summarized in the following short proposition.

"The money price of labor is necessarily regulated by two circumstances; the demand for labor, and the price of the necessaries and conveniences of life." (ibid.) The two factors in determining wages pointed out here are (1) the demand for labor (labor demand factor) and (2) the prices of necessities and convenience goods (consumer goods price factor). What, then, did Smith think of the relationship between these two factors? In other words, how did Smith conceived that these two factors were logically linked to determine wages? This point is summarized by Marx in the text quoted above, but here we will look at it a little more carefully, while implicit in Smith. For the time being, the following article will be helpful.

"The demand for labor, according as it happens to be increasing, stationary, or declining, or to require an increasing, stationary, or declining population, determines the quantity of the necessaries and conveniences of life which must be given to the laborer; and the money price of labor is determined by what is requisite for purchasing this quantity." (ibid.)

First, the labor demand factor that Smith is concerned with here is not its short-term fluctuations accompanying business cycles. Rather, the fluctuations are largely determined by the state of the long-term trend of the economy, whether the economy is developing like North America at the time, still like China, or stagnant like the East Indies. (ibid.) It was after Smith's death, in the second quarter of the 19th century that business cycles as the process of unfolding the internal contradictions of a capitalist society began to be seen in England. Smith's historical position is reflected in his ignoring the increase and decrease in demand for labor that accompanies the cycle.

For Smith, the long-term trend of labor demand determined in this way is the determinant of whether or not a further increase in the working-age population is necessary for the society concerned. This in turn determines the quantity of consumer goods appropriated to the worker. For example, if there is a need for an increase in the working population, they will be allocated more consumer goods. Wages are determined according to the prices required to purchase a certain amount of these consumer goods. Smith's wage determination theory can be schematized as follows: Long-term trends in labor demand (labor demand factor) → Whether or not population growth is necessary (labor supply factor) → Quantity of consumer goods (consumption material quantity factor) → Wages as prices of consumer goods (consumer goods price factors). If we look carefully based on this article, we can see that the labor supply factor and the consumption material quantity factor are hidden as intermediate terms between the labor demand factor and the consumer goods price factor pointed out earlier.

What are the characteristics of Smith's wage determination theory other than the above-mentioned tendency of labor demand? First, in comparison with Marx, it can be said that Smith is based on the same cost-of-subsistence theory of wage determination as Marx, in that he sees the quantity and prices of consumer goods as the basis of wages. However, as we just saw, Smith sees the amount of consumer goods workers receive as causally related to the speed of economic development. Therefore, Smith has a non-negligible difference from Marx, who regards the quantity of consumer goods as a product of the historical development stage of society. For Smith, on the other hand, no matter how developed a society might be, if that society were as stagnant as China was at the time, the amount of consumer goods received by workers would simply be few. For what Smith sees as a condition for high wages is rapid economic development. It is Marx, as we saw earlier, who has insight into the fact that the condition of "organic compositional constancy" is hidden here. This is also where the argument called "high-wage economy theory" was derived. According to "theory of high wage economy," as the rate of capital accumulation exceeds the population growth rate as the economy develops, a low profit rate coexists with a high profit amount and a high wage rate (Kobayashi 1977). For Marx, however, the quantity of consumer goods received by workers is the product of the absolute stage of development of society.[7] Therefore, no matter how stagnant a society may be, the higher the level of social development, the higher the wage level.

Secondly, Smith's and Marx's wage determination theories have a non-negligible difference in terms of the value of money that measures wages. For Smith, the price of consumer goods is generally understood as having the backing of invested labor, even though he is confused with the labor commanded theory.[8] On the other hand, the value of money, which measures

7 "the number and extent of his so-called necessary wants, as also the modes of satisfying them, are themselves the product of historical development, and depend therefore to a great extent on the degree of civilization of a country," (Marx 1867)

8 Smith mixes the labor invested theory and the labor commanded theory. In addition, as mentioned above, the value composition theory also coexists on the one hand. But in spite of this "confusion" of value definitions, Smith finds, in

the price of consumer goods, is said to be determined by the quantity theory of money, that is, by the relative relationship with the quantity of circulating commodities.[9] In this respect, Smith is very different from Marx, who sees money as one of the commodities and sees the amount of labor behind its value. Even though Smith criticized the bullionism that distinguishes money from general commodities, Marx asserted that it still exists in Smith. (Marx 1862-1863)

Thirdly, in Smith's theory of wage determination, the perception of labor supply is also a long-term trend, corresponding to the long-term trend of labor demand. "The Wealth of Nations" elaborates on the labor supply theory later in Chapter 8 as follows.

"If this (labor-quoter) demand is continually increasing, the reward of labor must necessarily encourage in such a manner the marriage and multiplication of laborers, as may enable them to supply that continually increasing demand by a continually increasing population. If the reward should at any time be less than what was requisite for this purpose, the deficiency of hands would soon raise it; and if it should at any time be more, their excessive multiplication would soon lower it to this necessary rate." (Smith 1776)

---

effect, "wherever he develops his argument, a correct definition of the exchange-value of commodities—that is, the determination of value by the amount of labor or working hours— " (Marx 1862-1863) Furthermore, "in his work on the natural price of wages, A. Smith practically escapes—at least in places—to the correct value determination of commodities." (ibid.)

9 However, Smith (1776) also has the following description, which is not necessarily consistent with the quantity theory of money. Smith also has a theory of money based on the invested labor theory of value and the labor commanded theory. "The discovery of the abundant mines of America reduced, in the sixteenth century, the value of gold and silver in Europe to about a third of what it had been before. As it costs less labor to bring those metals from the mine to the market, so when they were brought thither they could purchase or command less labor." Kobayashi (1977) discusses the turbulence of Smith's monetary theory.

For Smith, labor supply is the dependent variable of labor demand accompanying economic growth. Economic growth increases the demand for labor and raises wages, which promotes the reproduction of workers and increases the working population. If wage increases are not sufficient, the reproduction of workers will be restrained and the increase in the working population will be restrained, resulting in a labor shortage that will raise wages. Conversely, if wage increases are excessive, excess labor will push wages down. Since the demand for labor, which is the basis of this movement, has been grasped as a long-term trend, the supply of labor, which is ultimately determined by that demand through wages, has also been grasped as a long-term trend. From the above, according to Smith's theory of wages, the cost of generational reproduction of labor-power must be included in wages. This point was later taken over by Ricardo[10] and Marx's theory of wages.

This is the end of examination of Smith's wage determination theory based on the long-term trend of labor demand. On that basis, let us see how Smith explains the previous problem, namely the wage gap between urban and remote areas. On this point, "The Wealth of Nations" has the following description in Chapter 9 of Part 1, "on the profits of stock."

"In a thriving town the people who have great stocks to employ frequently cannot get the number of workmen they want, and therefore bid against one another in order to get as many as they can, which raises the wages of labor, and lowers the profits of stock. In the remote parts of the country there is frequently not stock sufficient to employ all the people, who therefore bid against one another in order to get employment, which lowers the wages of labor and raises the profits of stock." (Smith 1776)

---

10 "The natural price of labor, therefore, depends on the price of the food, necessaries, and conveniences required for the support of the laborer and his family." (Ricardo 1817)

Smith says, 〈high capital accumulation → large labor demand → relatively high wages〉 in urban areas, and 〈low capital accumulation → low labor demand → relatively low wages〉 in remote areas. Here, we can see that the long-term trend theory of wage determination that we saw earlier is used to explain regional disparities in wages. In addition, the difficulty of geographical movement of labor compared to general commodities ("human mobility difficulties") is a precondition. This is because if there is free labor movement between cities and remote areas, such wage disparities will disappear in a relatively short period of time through labor movement. However, as I mentioned earlier, a more in-depth examination of the content of "human mobility difficulties" is necessary.

Therefore, in order to examine the content of "human immobility", let us refer to the description in Part 1, Chapter 10, "on wages and profit in the different employments of labor and stock." At the beginning of the chapter, while stating "the whole of the advantages and disadvantages of the different employments of labor and stock must, in the same neighborhood, be either perfectly equal or continually tending to equality," the chapter considers five circumstances in which one occupation compensates for the small monetary gain and another occupation offsets the large monetary gain. The chapter shows that the equality between occupations is realized in terms of advantages and disadvantages as a whole. In addition, the chapter points out three conditions for such equality to occur, but let us look here at the third point, which is related to the problem at hand. That is, "the employments must be the sole or principal employments of those who occupy them." (ibid.)

"When a person derives his subsistence from one employment, which does not occupy the greater part of his time, in the intervals of his leisure he is often willing to work as another for less wages than would otherwise suit the nature of the employment.

There still subsists in many parts of Scotland a set of people called Cotters or Cottagers, though they were more frequent some years ago than they are now. They are a sort of outservants of the landlords and farmers. The usual reward which they receive from their masters is a house, a small garden for pot-herbs, as much grass as will feed a cow, and, perhaps, an acre or two of bad arable land. When their master has occasion for their labor, he gives them, besides, two pecks of oatmeal a week, worth about sixteen pence sterling. During a great part of the year he has little or no occasion for their labor, and the cultivation of their own little possession is not sufficient to occupy the time which is left at their own disposal." (ibid.)

At that time, the surviving "Cottagers" in Scotland received (1) a house and a farmland (including a vegetable garden and a pasture) and (2) oatmeal in kind as remuneration for their labor. Since they were tenant farms, their fixation to the land was not as strong as in the case of owner farmers, but they were characterized by their sedentariness compared to landless workers. Therefore, the existence of "Cottagers" was one of the factors that determined "difficulty in moving people." They were also workers who worked for lower wages than they should have received. Therefore, the prevalence of "Cottagers" in Scotland had an effect on the rural wage gap between Scotland and England. In the first place, Smith faced a society in which independent producers still existed widely, and even if there were workers, it was a society in which they could frequently revert to being independent producers. Smith says: "In years of plenty, servants frequently leave their masters, and trust their subsistence to what they can make by their own industry." (ibid.) In other words, it was a society in which a class completely separated from the land, who were constantly forced to sell their labor-power for a living, was not yet fully formed.

Furthermore, "immobility of people" was artificially created by the policies that formed the basis of the labor organization in England during the Mercantilist period, namely, the corporation laws and the poor laws (Polanyi 1944). In connection with regional wage disparities, it is important to note that in Europe, seven years has been established as the normal period of apprenticeship since ancient times (Smith 1776). In Scotland, however, there was no country in Europe where the corporation laws were less oppressive than in Scotland, such as the usual three-year apprenticeship period for even the most sophisticated professions. And this has resulted in lower wages for skilled urban workers in Scotland. Regarding the poor laws, "The Wealth of Nations" points out the special circumstances of England as follows.

"It consists in the difficulty which a poor man finds in obtaining a settlement, or even in being allowed to exercise his industry in any parish but that to which he belongs. It is the labor of artificers and manufacturers only of which the free circulation is obstructed by corporation laws. The difficulty of obtaining settlements obstructs even that of common labor." (ibid.)

"The very unequal price of labor which we frequently find in England in places at no great distance from one another is probably owing to the obstruction which the law of settlements gives to a poor man who would carry his industry from one parish to another without a certificate." "The scarcity of hands in one parish, therefore, cannot always be relieved by their superabundance in another, as it is constantly in Scotland, and, I believe, in all other countries where there is no difficulty of settlement. In such countries, though wages may sometimes rise a little in the neighborhood of a great town, or wherever else there is an extraordinary demand for labor, and sink gradually as the distance from such places increases, till they fall back to the common rate of the country; yet we never meet with those sudden and unaccountable differences in the wages of neighboring places which we sometimes find in England, where it is often more difficult for a poor man to pass the artificial boundary of a parish than an arm of the sea or a ridge of high mountains, natural boundaries which sometimes separate very distinctly different rates of wages in other countries." (ibid.)

In Scotland, as we saw earlier, the "immobility of humans" caused by their fixation to the land is stronger than in England. This, in the first place, reduced wages in rural Scotland. If we focus only on this point, the disparity between urban and rural wages for unskilled labor should be more conspicuous in Scotland. In reality, however, the "sudden and unaccountable differences in the wages of neighboring places" was a social phenomenon more prominent in England than in Scotland (**Table 1-1**). To explain this point, Smith sought the restriction of movement of the poor by the poor law, which is unique to England.

The remaining problem in **Table 1-1** is the relative low wage in urban area in Scotland, but Smith did not give a sufficient explanation for this point. In this regard, it is necessary to take into account a different problem from "immobility of labor" and low wages for "Cottagers," namely, the disparity in the speed of economic development in cities between England and Scotland. Edinburgh's economic development was slower than that of London, and the movement of unskilled labor, unimpeded by the poor law, brought about the spread of low wage from rural area to urban area. Therefore, in Edinburgh, wages were relatively low.

4. Regional Disparities of Wages in "Capital"

How is the issue of inter-regional wage disparity dealt with in "Capital?" First, as a matter of fact, "Capital" does not include the recognition of regional wage disparity in its logic. There is a good reason for this that is connected with the methodology of "Capital." This is because, if the recognition of regional wage disparity was incorporated into the basic logic, the character of the

"Capital" would be greatly different from the current one. This can be seen by conducting a thought experiment to see what kind of results would arise if there was regional wage disparity. If we assume the existence of inter-regional disparity in wages, we must acknowledge the existence of inter-regional disparity in the rate of surplus-value. (The working hours shall be a given uniform at the same point in time.) However, there is, of course, no such discussion of regionality in "Capital." If we accept the existence of regional disparity in the rate of surplus-value, we cannot derive the average rate of profit, which is based on the existence of a general rate of surplus-value, as in "Capital." And if there is no average rate of profit, the theory of rent, which is based on it, cannot be developed like "Capital." In short, if we accept the existence of regional wage disparity, we cannot derive the major economic categories developed in "Capital" in the current way, as indicated by the above brief description. Please understand that we are discussing wages for unskilled labor here as well.

What Marx actually assumes in "Capital" is the opposite. In other words, there is a tacit understanding of the universal equality of wage levels among regions, and the existence of uniformity in the labor market that makes this possible is assumed. Assuming the uniformity of the labor market means the existence of rapidly moving capital and workers in order to eliminate differences in wage levels between regions in a short period of time. Since the portion of capital invested in fixed capital is relatively difficult to move, the free movement of labor is of great importance in order to realize uniformity in the labor market. Thus, the image of the workers assumed in "Capital" is, first of all, freed from legal restrictions on movement. But that alone is not enough. Second, they are "free as birds," separated from the means of production (Marx 1867). Third, they are unskilled workers with low barriers to inter-occupational mobility (ibid.).

5. Implications of the Worker Image to the Theory of Capital Accumulation

From the above, it is clear that the images of workers envisioned in "The Wealth of Nations" and "Capital" differ greatly. Smith's workers are land-bound, artificially restricted in movement, and highly skilled. Marx's workers are unskilled laborers who are separated from their land and whose artificial restrictions on movement have been abolished. One of the logical consequences is the contrast between the two in the regional wage disparities. What, then, does this difference in worker images bring about in terms of accumulation theory? We will look at this point below.

The theory of accumulation in Chapters 21 to 23 of Volume 1 of "Capital" extends from the theory of simple reproduction as the reproduction process of class relations in the capital system to the theory of reproduction process on an enlarged scale (conversion of surplus-value to capital = accumulation). The latter theory of accumulation develops the theory of relative surplus-population as a law for securing the labor force in an established capitalist society.

By the way, in order for the labor force to be procured through an increase in the organic composition of capital, the existence of workers who are free in a double sense (the *Vogelfrei* property) is a prerequisite. This does not only mean in the general sense that labor-power must be freed from feudal bondage and must have lost its land (means of production) before it can be mobilized to capital. Another point is that if peasants (the non-capitalist sector), freed from feudal bondage, are the source of labor-power, labor-power can easily be procured from such sources without the capitalist population law functioning. In other words, the creation of a relative surplus-population through the increase of the organic composition of capital, that is, the capitalist labor procurement mechanism, does not need to function. As long as the capitalist sector can employ the abundant labor force supplied by the non-capitalist sector at low wages, the capitalist logic of restricting the introduction and use of labor-saving technology works. As a result, the operation of the capitalist labor procurement mechanism by creating a relative surplus-population is suppressed. Using the development economics theory, this supply of labor was initially carried out as the so-called unlimited supply of labor of Lewis (1954), and after the turning point was passed, it was accompanied by an increase in agricultural labor productivity. As a precondition for the functioning of the capitalist population law, it is necessary that the supply of labor from these non-capitalist sources be exhausted, whatever the natural increase in population there may be. Because of the existence of capitalist restrictions on the introduction and use of such labor-saving technologies from the viewpoint of cost-effectiveness, theoretically, the theory of primitive accumulation, whose main content is the separation of the means of production and the producers, is the theory of the creation process of the preconditions of the theory of relative surplus-population.

From the inferences thus far, the reason why "The Wealth of Nations" assumed an accumulation of unchanging organic composition and could not develop a theory of relative surplus-population was that the image of workers assumed therein did not have the *Vogelfrei* characteristic. On the other hand, "Capital" was premised on the *Vogelfrei* characteristic of workers, so it could to develop the theory of relative surplus-population.

Looking back at what I have just said from the viewpoint of the theoretical significance of the theory of primitive accumulation in "Capital," the theory reveals the *Vogelfrei* characteristic of workers, which is the precondition for the functioning of the capitalist law of population. However, the *Vogelfrei* characteristic of workers is a matter of historical fact whether it exists or not. Therefore, the theory of primitive accumulation can only be developed as a historical description. However, the existence of *Vogelfrei* workers, which is a historical fact, has a theoretical significance as a precondition for the functioning of the law of capitalist population. In this way, the primitive accumulation theory supplements the theory of relative surplus-population. The primitive accumulation theory is in Chapter 24 of "Capital," Volume 1, accordingly it is located immediately after Chapter 23, which develops the theory of relative surplus-population. The primitive accumulation theory, which is a historical description, is by no means a superfluous theory. It has the theoretical significance of mediating the transition from accumulation with organic compositional invariance to accumulation accompanied by increase of organic composition.

There must, therefore, be two themes of such a primitive accumulation theory. One is the depiction of the abolition of artificial movement restrictions that impede the free movement of labor-power. And the other is the depiction of the dissolution of the regional fixation of humans who were tied to the land. However, Marx regards the latter, "the historical process of divorcing the producer from the means of production," as the most essential element of primitive accumulation (Marx 1867). This is probably because this aspect characterizes primitive accumulation as the process of creating workers as a class that cannot survive without selling its own labor-power. In terms of agriculture, this is a two-step process in which the community that has protected the farmers until then is destroyed through legal and illegal means, and the peasant classes is dissolved. This is a great contrast to the fact that in the past even rural workers had rights to communal land.

Workers who lost their ties to the land initially stayed in rural areas and formed a surplus-population there. However, they gradually transformed into urban workers while repeating wave-like movements (Yoshioka 1981). The mule spinning machine, which was introduced in the 1790s, established mule factories based on the use of steam engines, and concentrated the cotton spinning industry in cities with favorable production conditions. (ibid.) [11] However, according to Mochida (1996), in the "first half of the 19th century," the workers created as a result of the differentiation of the peasant class remained in farming villages "except in the vicinity of industrial areas," forming a rural surplus-population. However, "in the 1840s, real industrial wages began to rise," "and by 1850, when the railroad network was completed, the outflow of the rural population of southern England began." Furthermore, "the money wages of agricultural workers rose by 40% from the early 1850s to the early 1870s," and "with the completion of the national unified market, the surplus labor force that remained in rural areas in the first half of the 19th century was wiped out." (ibid.) Mochida's argument about the migration of labor from rural areas to cities is almost consistent with Ferrand's speech to the British House of Commons in 1863 regarding labor shortages during the economic boom, which is quoted in Chapter 8 of Volume 1 of "Capital." That is, during the industrial labor shortage in the economic boom of 1834, there was still a supply of labor from rural areas to urban areas (existence of rural surplus-population). But in the economic boom of 1860 it was no longer possible (depletion of rural surplus-population). Along with the exhaustion of the rural surplus-population,[12] the factory owners petitioned the workhouses for an

11 "Except London, there was at the beginning of the 19th century no single town in England of 100,000 inhabitants. Only five had more than 50,000. Now there are 28 towns with more than 50,000 inhabitants." (Marx 1867)

12 However, Chapter 8 of "Capital," Vol. 1, in a note, gives a more comprehensive explanation of the labor shortage in 1860 (Marx 1867). In other words, the causes of the shortage are (1) the population decline in Ireland, (2) immigration from farming areas in England and Scotland to Australia and America, and (3) the clear decline in the population of some rural areas in England. It then cites the "draining of labor from farm villages to cities" as a factor in explaining (3). The note further mentions the destruction of life force by overwork. In connection

alternative supply of labor consisting of poor children and orphans (Marx 1867). [13] The year 1834, the starting point of Marx's description, was also the year in which the poor law was revised in England, the level of poverty relief

---

with (2), in the middle of the 19th century, immigration from England to North America and Australasia increased rapidly. On this point, Morita (1997) argues that "the overwhelming majority of immigrants from England were actually Irish," and that "they were immigrants from the colonized 'periphery' who were incorporated into the vertical division of labor led by the British capitalist movement. (Ireland was annexed to the United Kingdom in 1801.)" Morita emphasizes this point "to dispel the false notion that the expression 'outbound immigration from England' can create." This is because "immigrants from England" tend to conjure up the impression that people were migrating from the "core," where capitalist industrialization has been achieved through the Industrial Revolution, to the non-industrial "periphery," which was still capitalistically undeveloped. In general, Morita also states that "a considerable portion of 19th-century European immigration was not from within the capitalist sector, but rather from the labor force expelled from the 'periphery' that had to be dismantled and reorganized in the process of permeation and integration of capitalism." In relation to (3), Broadberry et.al. (2015: p.367) report that the increase (annual rate) of labor productivity by sector in Britain between 1801 and 1851 was 1.23% for industry and 0.71% for services, but only 0.10% for agriculture, which was far short of the annual increase of 0.74% for agricultural output. This would mean that a natural increase of 0.64% per annum in the agricultural labor force (laborers and farm household members) was necessary despite the labor migration from rural areas to urban areas. But this would not be a rapid increase in absolute terms.

13 In the latter half of the 19th century, rural workers were also subsumed under capitalist agriculture, but many of them still seemed to be tied to the land. Looking at land distribution in England and Wales in the so-called "New Domesday Book" of 1873, 87 percent of the land was concentrated in the hands of 4 percent of the landowners, while 73 percent of the cottagers owned less than an acre. "Most of them, including agricultural workers, were small-scale farmers who had their main business other than farm management and managed to earn part of their household income from this small land." (Ouchi 1977) However, although the latter half of the 19th century was a period of expansion of capitalist farm management in England and Wales, the number of agricultural workers actually decreased. This shows the peculiarity of capitalist development in agriculture, and the increase of the organic composition accompanying the development of agricultural productivity brought about not only a relative decrease in variable capital but also an absolute decrease. (ibid.)

was lowered, the parishes were consolidated, and the central administrative body (poor law commission) was established. It was also the year in which interregional labor migration became active thereafter (Yoshioka 1981).

On the other hand, in the process of escaping from the recession after the panic of 1825, the introduction of new production methods led to the raise of organic composition, and the urban surplus-population, not the rural surplus-population, began to emerge from the movement of capital accumulation.[14] In 1825, the self-actor was invented, an improved version of the mule spinning machine. As a result, the problems of unemployment and poverty became more serious, and the labor movement became sharper. (Hudson 1992) In other words, the formation of the "National Trade Union Grand Confederation" led by Owen (1834) and the development of the "10-hour movement" instigated by Osler were seen. (Yoshioka 1981) It was also the time when Wakefield developed a discussion of "organized immigration" in "England and America" (1833), and the colonies came into the spotlight as an outlet for alleviating the problems of surplus-population and poverty. (Yoshioka 1981) From the above, it can be said that the 1830s was a turning point in England, when the surplus-population of the peasants was exhausted, the formation of a modern working class with no connection to the land, and the establishment of the law of capitalist population through the increase of the organic composition accompanying the business cycle.

6. Conclusion — Diversity of Primitive Accumulation

If, as in this chapter, the primitive accumulation theory is regarded as the

---

14 The shift to large-scale mechanized industry in England began around 1760 as a countermeasure against unrestrained skilled workers. On the other hand, the adoption of new technology in the 1820s, which is the subject of this text, was aimed at improving labor productivity during the recession during the economic cycle. Both have different motives.

definition and development of the *Vogelfrei* of the working class as a precondition for the theory of relative surplus-population, the primitive accumulation as an actual historical process (not the primitive accumulation theory) is the formation process of such a *Vogelfrei* property. This conception of primitive accumulation differs from the historical starting point of capitalist society, that is, the image of primitive accumulation as the process of transition from feudalism to capitalism (Otsuka 1956a, Uno 1936). Indeed, as I mentioned at the beginning of this chapter, the process of forming the *Vogelfrei* nature of the working class in England corresponded to the period of social transition from feudalism to capitalism. This was not the case in countries and regions other than Great Britain. These countries and regions became capitalist more or less under pressure from the first country. In the latter, new capitalist productive forces with a high degree of organic composition were introduced exogenously by grafting onto the old relations of production. In that case, even after the capitalist society was formed, and even after the capitalist reproduction trajectory was established without waiting for the disintegration of the non-capitalist sector, the capitalist sector was supplied with labor from the non-capitalist sector. During that time, the triggering of the labor supply mechanism by the creation of a relative surplus-population through the increase of the organic composition of capital was restrained. For this reason, in latecomer capitalist countries, as mentioned in Section 1 of this chapter, the integral character of the various factors involved in the formation of the capitalist society is lost. Therefore, it is necessary to grasp each factor analytically. The following three are conceptually distinguished. (1) "Transition of society from feudalism to capitalism," consisting of the modernization of land ownership; (2) "Formation of a capitalist society" consisting of the establishment of an industrial structure of two sectors; (3) "Establishment of a capitalist society" from the perspective of the formation of the capitalist population law. Therefore, it is thought that in late-starter capitalist countries, the primitive accumulation continues even after the formation of the capitalist society. How is the end point of primitive accumulation defined? That is when the capitalist sector has finished draining almost all of the labor force from the non-capitalist sector. It was also when capital had to establish a capitalist procurement mechanism for the labor force through the creation of a relative surplus-population through the business cycle and the increase of organic composition. It happened to be in England at the beginning of the nineteenth century, and from this came the understanding that the primitive accumulation was confined to the theory of transition from feudalism to capitalism.

Chapter 2 : Starting Point and Ending Point of Primitive Accumulation

1. Theme

The image of so-called primitive accumulation in economics is reflected in the famous phrase "capital comes dripping from head to foot, from every pore, with blood and dirt." in "Capital," Volume 1, Chapter 24, "So-called Primitive Accumulation" (Marx 1867). Roughly speaking, the two axes of logic are the separation of the rural population from the land, which is the basis of the entire historical process of primitive accumulation (issue 1: essential element), and the process of formation of the first capitalists at the other extreme (issue 2). In them are incorporated descriptions of the purification of the working masses as wage workers and the historical tendency of capitalist accumulation through the dialectic of "denial of denial." However, a coherent understanding of the content of Chapter 24 is not easy, as we will see below.

For example, even with regard to the main points of Chapter 24 (Issues 1 and 2), there are some unorganized parts. With regard to point 1, Marx pays attention to the 1) "gradual expansion of the exploitation of wage labor" by small capitalists, while at the same time he puts to the fore the 2) "sudden and violent proletarianization of large groups of people." In relation to issue 2, he cites two groups of capitalists as origins; (a) independent small producers of commodities, and (b) early merchant capital and usury capital.

Against this background, which can be described as the "confusion" of these descriptions, two trends have arisen in Japan regarding the understanding of primitive accumulation. According to Otsuka, the "decomposition" through the productivity competition of the "middle class of producers" gave birth to "industrial capitalists" and workers (Otsuka 1956a). In other words, the image of primitive accumulation of Otsuka is a combination of the above 1) and (a) that can be described as idyllic. The Uno school, on the other hand, regards the accumulation of merchant capital and loan shark capital as primary, and emphasizes the significance of violent processes in the creation of the proletariat (Uno 1936). In other words, it is a combination of 2) and (b).

This controversy in Japan narrowed down the theme to grasping the formation process of capitalist society in England, and this controversy was significant in that it helped deepen the understanding of this theme in Japan. However, as will be seen later in Section 4, according to Marx's later images, which became clear through examination of the French version of "Capital" (Marx 1872-1875) and his letter to Zasurich (Marx 1881), primitive accumulation is diverse by country. Considering this point, limiting the discussion of primitive accumulation to the formation process of the British capitalist society is rather narrowly restricting the problem.

As recognition of this extensional diversity of primitive accumulation deepens, what must be asked again is what is primitive accumulation in the first place, or in other words, the question of its connotations.

By the way, Mochizuki (1977) has a description of the formation process of Marx's primitive accumulation theory, which may give us a hint when considering this point. Let's quote it below.

"Marx first developed the primitive accumulation theory in the 'chapter on capital' in the 1857-59 draft '*Grundrisse*.' This is the series of essays, now published under the title 'Pre-Capitalist Economic Formations' (abbreviated 'Formations'), consisting of two sections according to Marx's own classification. In fact, just before these two sections there is a rather theoretical section titled 'primitive accumulation', and it is necessary to grasp them including it. Marx wrote this series of statements as part of his theory of capital accumulation, which takes the form of capital circuit = turnover theory in the '*Grundrisse*.'

According to the draft plan of 1859 (actually 1861), he wrote 'Formations' as Chapter 4, 'primitive accumulation' of 'Criticism of economics.' If so, the part that discusses the three communities, Asian, Classical, and Germanic, should be read as an introduction to the theory of primitive accumulation."

This Mochizuki's assertion that "Formations" must be read as an introduction to the theory of primitive accumulation is very suggestive to me. However, from this point on, we must not only point out this point, but also go one step further and ask what kind of significance there is in reading "Formations" as an introduction to the theory of primitive accumulation. Therefore, in this chapter, we will consider the content of the concept of primitive accumulation while examining the relationship between "Formations" and the theory of primitive accumulation. In the previous chapter, I examined the relationship between the theory of capital accumulation (Chapter 23, Capital Vol. 1) and the theory of primitive accumulation. I argued that the primitive accumulation should be understood as the process of forming the preconditions for the establishment of capitalist mechanisms of labor force creation. This was, so to speak, related to the "end" of the primitive accumulation process. On the other hand, what is clarified in this chapter will be related to the "beginning" of the primitive accumulation. As long as the primitive accumulation is a historical process, it naturally has a "beginning" and an "end."

2. "Principle of Survival" in the Community

What kind of content should we read from "Formations," which is known for their obscure writings? Regarding "Formations," there is a classical interpretation by Otsuka (1955) that emphasizes the form of community ownership, which has had a profound impact on the understanding of community in Japan. Certainly, on the one hand, Marx's awareness of the problem in "Formations" posits the capitalist society as the pole of society in which private ownership prevails. As a logical development sequence leading up to it, various forms of Asian, Roman, and Germanic communities are positioned. While respecting Marx's awareness of the problem, it is quite natural that the description of the draft "Formations" should crystallize into the Otsuka community theory. The Otsuka community theory focuses on the "inherent duality" of communal occupation and private occupation that exist within the community. However, if we take a step back from this question and think about how community theory should be in the first place, it is clear that community theory cannot be reduced to ownership form theory. This corresponds to the inability to reduce the theory of capital to the theory of capitalist property. Just as capitalism requires a theory of internal laws of motion developed on the basis of capitalist property relations, community theory needs to elucidate its internal logic on the basis of communal property relations. This very point must constitute one of the themes of the theory of community alongside the theory of property forms. Here, let us examine the description of "Formations" from this point of view. It is believed that this is more useful in understanding what is actually denied by the primitive accumulation. In fact, in spite of its basic character as a theory of property form, "Formations" contains useful descriptions for gaining points of view when clarifying the internal logic of the community.[15] For example, the following sentence:

"in all these forms (Asian, Roman, Germanic forms of property — quoted), where landed property and agriculture form the basis of the economic order, and consequently the economic object is the production of use-values - i.e., the *reproduction of the individual* in certain definite relationships to his community, of which it forms the basis:" (Marx 1858)

---

15 A more detailed discussion of this point can be found in Chapter 4 of this book.

According to this writing, the economic activity common to the three forms of community is the production and distribution of self-sufficient goods, mainly agricultural products, with the aim of ensuring the survival of community members and their reproduction across generations. This is a completely different society than a capitalist society in which economic activity aims at maximization of economic value. Therefore, I shall refer to the purpose of such a community as the "principle of survival" below. Collective ownership of land and the stockpile of wealth and its redistribution within communities serve this purpose. These have sometimes been called "community regulations," but they are nothing more than means for the "principle of survival." There are laws (customs) that correspond to collective ownership of land and the function of stockpiling and redistributing wealth within the community. Furthermore, in order for these laws (customs) to function stably, religious-ideological forms are necessary to support them.

In addition, as long as a pre-capitalist class society is established on the basis of the community, it must be consistent with the "principle of survival." As long as it is a class society, it is inevitable that part of the surplus-value generated within the community belongs to a specific group that reigns as the ruling class. On the one hand, the attribution is justified as compensation for the performance by that particular group of functions universally found in the community, including the stockpiling of wealth and its redistribution among its members.

3. Primitive Accumulation as Denial of the "Principle of Survival"

What implications, then, does understanding the community from the perspective of the "principle of survival" have in understanding the primitive accumulation? And what aspect of the primitive accumulation does this illuminate? In thinking about this problem, it would be effective to consider how a society based on the "principle of survival" would be undermined. This is because, in the process of this collapse, there must be elements that must be epoch-making as primitive accumulation.

Within a community that has a "principle of survival" and corresponding legal and ideological forms, initially "pursuit of private interests" is such that it may conflict with the right to exist of others. Therefore, it appears as an evil = illegal act that should be denounced in the society concerned. Sharing is incompatible with the pursuit of personal gain. The usurpation of communal land would be the most unlawful act. However, what was initially regarded as an illegal act gradually acquires social legitimacy as long as the act embodies the development of productive forces. Then the "pursuit of private profit" and its corresponding legal consciousness (private land ownership and private occupation of products) and ideological forms grow. Since the community continued to suffer from starvation due to low productivity, the development of productivity was permitted by the "principle of survival." However, as long as the process of development of productive forces is accompanied by winners and losers through competition, it ultimately comes into conflict with the "principle of survival."

On this point, let us first look at the example of England, where the primitive accumulation is said to have developed "in the classic form" (Marx 1867). There, the technological basis for capitalist farm management was created through a series of agricultural technology improvements since the 1730s. Such technologies consist of the technique of stripe sowing and cultivating of Tull, the Norfolk four-course system[16] based on "fodder turnip - spring barley - red clover - winter wheat," and Bakewell's livestock breeding. This series of technologies

16 Tsuge (2010) for the Norfolk four-course system. As for the evaluation made by Marx and Liebig that the development of such agricultural technologies is nothing more than the development of "sophisticated predatory agriculture," see Shiina (1976).

brought about the parallel progress of grain farming and livestock farming. Smith (1776) emphasizes the benefits of "enclosure" in pasture: "The advantage of enclosure is greater for pasture than for corn. It saves the labor of guarding the cattle, which feed better, too, when they are not liable to be disturbed by their keeper or his dog." Therefore, landowners actively carried out "enclosure," expecting more benefit (land rent) from capitalist farm management (Shiina 1962). These two aspects, (1) establishment of capitalist farm management and (2) abandonment of the old land system, i.e. establishment of modern land ownership, constitute the content of the agricultural revolution.

As a result of the above, a conflict arose between the community's "principle of survival" and the new way of life based on the "pursuit of private interests" on the other side, but it could not have been a peaceful process. This is because both sides argued that they had legitimate rights to each other. One had a legal and ideological system aimed at the coexistence of community members, and the other had a legal and ideological system aimed at the development of productivity and the "pursuit of private interests." Locke (1690), representing the latter position, states as follows: "As much land as a man tills, plants, improves, cultivates, and can use the product of, so much is his property. He by his labor does, as it were, inclose it from the common." He encourages "enclosure," with the condition that it be based on self-labor. However, when legitimate rights from each stand collide, it is private and state violence by the prevailing class that determines the outcome. Violence was necessary to break the community.

"Capital," Volume 1, Chapter 24 is full of examples of violent resolution of the conflict between the "pursuit of private interests" as basis for a new way of life and the community's "principle of survival." Let's start from the beginning to emphasize the contrast with the period after the agricultural revolution.

a) Late 15th century to early 16th century: The rapid rise of Flemish wool manufacturers, and the corresponding rise in the price of wool in England. As a result, "the great feudal lords created an incomparably larger proletariat by the forcible driving of the peasantry from the land, to which the latter had the same feudal right as the lord himself, and by the usurpation of the common lands." (Marx 1867) In other words, along with the first enclosure movement, the transformation of arable land into sheep-walks progressed. In those days, however, the looting of common land was performed as a personal assault. Legislation was "terrified" at this revolution and sought to prevent land grabbing from peasants by issuing an enclosure ban, banning the conversion of arable land into sheep ranching and ordering the restoration of peasant land and buildings. At this point the state still had sympathy for the feudal rights of the peasants.[17]

b) 16th century: The Protestant Reformation and the consequent colossal spoliation of the church property. At that time the Catholic church was the feudal proprietor of a great part of the England land. "The estates of the church were to a large extent given away to rapacious royal favourites, or sold at a nominal price to speculating farmers and citizens, who drove out, *en masse*, the hereditary sub-tenants and threw their holdings into one." This created a mass pauper, but the legally guaranteed property of the poor folk in a part of the church's tithes was tacitly confiscated.

c) Late 17th century: Abolition of the feudal tenure of land after the restoration of the Stuarts. The landed proprietors vindicated for themselves the rights of modern private property in estates to which they had only a feudal title. In addition, state lands were stolen after the Glorious Revolution. As a result, the land became a freely traded commodity. The domain of

---

17 Huberman (1936) noted that anti-enclosure laws were first issued in 1489 and were then issued repeatedly throughout the sixteenth century. However, the frequent occurrence of such laws indicates that they were largely ignored.

modern agriculture on the large farm-system has expanded.

d) 18th century: The law itself becomes now the instrument of the theft of the people's land. That is, the start of secondary enclosure movement. In the 18th century, parliamentary permission to "enclosure" petitions was primarily granted in the form of Private Acts. However, in 1801, the General Enclosure Acts were enacted to govern all "enclosures" with a single law, referring to cases of "enclosures" that had been practiced in various places until then. "Its establishment signified the final stage in the history of the enclosure, the state of affairs at the time, in which there could no longer be any agricultural development without enclosure, and the eventual disappearance of open arable land in England." (Shiina 1962)

e) Land clearing in the Scottish Highlands. The inhabitants of it, the Celts, were organized in clans, each of which was the owner of the land on which it was settled. The representative of the clan, its chief or "great man," was only the titular owner of this property. The "great man" , on his own authority, transformed his nominal right into a right of private property, and as this brought him into collision with his clansmen, resolved to drive them out by open force. An example of this in the first half of the nineteenth century, the "clearing" of the Duchess of Sutherland, shows the destruction of villages in order to make use of the whole county, already reduced to 15,000 inhabitants, as rented sheep-walk, converting all the cultivated land into pasturage. Later, part of the sheep-walks was turned into deer preserves.

From the above descriptions a) to e), the assertion of "pursuit of private profit" gradually grew and earned social legitimacy while opposing the "principle of survival" of the community. The former eventually earned legislative backing. However, why did things move in this way? In this regard, Shiina (1962) describes the peasants' opposition to the "enclosure" of the Second Enclosure period in England and the changes in it as follows. He describes the circumstances in which the growing differentiation and dissolution of the peasant classes along with the development of agricultural productivity turned the opposition to the "enclosure" of communal land into a mere formality.

"Opponents to the second enclosure movement were primarily the peasant classes dependent on common land. But the peasants were already disintegrating to the point where they did not form a single class. In addition, some capitalist land-renting farmers and small-scale farmers who were starting to quit farming, did not actively support the opponents. As a result, the movement against enclosure did not become so large." (ibid.)

According to Lefebvre (1939), in the latter half of the 18th century in France, the royal power authorized the freedom of "enclosure" and the distribution of common land in some provinces. At this time, there was a growing public interest in British-style farm management. During the French Revolution, however, peasant criticism of this infringement of joint ownership of land increased. At that time, the peasant class in France was no longer monolithic, but its differentiation and decomposition has not progressed as much as in England. Large farm management profited from "enclosure" and division of common land, but there were many peasants who criticized it. The rural movement in 1789 to restore common land ownership and abolish feudal tribute resulted in a general rebellion by the peasants, which destroyed the "enclosure" everywhere and restored communal grazing. The peasants also reclaimed the communal lands that had been taken from them. This success was due to the circumstances described above.[18]

---

18 Therefore, in Victor Hugo's historical novel Quatrevingt-treize (1874), which depicts the climax of the French Revolution, Gauvain, a young commander of the republican army who went to suppress the royalist rebel army, says: "Three-quarters of the soil is waste land; clear up France. Put an end to useless pastures, divide the communal lands. Let every man have a piece of ground, and

Now, if we regard the primitive accumulation as a historical process in this way, the significance of the controversy over whether its main aspect is a violent process or an idyllic process must be relativized. The primitive accumulation contains as an important element the denial of the community-based "principle of survival." It is a denial of the way society should be for the purpose of coexistence, although the members of the community are poor. In order to deny a society whose purpose is the "principle of survival," it is necessary for the activities and consciousness aimed at "pursuing private interests" to grow gradually within the community (pastoral process). These activities are initially carried out illegally and run into conflict with the "principle of survival" (violent process). However, when they embody new and advanced ways of productive power, they gradually acquire legitimacy (pastoral process). It should also be noted that the social class that embodied the "principle of survival" is hollowing out due to the progress of stratification in society, and the foundation on which opposing forces rely is being undermined. When they finally reach the point where they collide with the "principle of survival" in order to finally destroy the common land, the conflict between the two reaches a climax (violent process). In this way, the primitive accumulation is a historical process woven by the intermediary and dialectical intertwining of pastoral and violent processes. In other words, primitive accumulation is an idyllic process and at the same time a violent process. By interpreting the process of divorcing the producer from the means of production (Marx 1867), which constitutes the essential element of primitive accumulation, as a "social revolution" in the sense of denying society with the purpose of "the principle of survival," the above two aspects can be understood within a single logic. Polanyi (1944) shares this idea of seeing a great

---

every piece of ground have a man. It would multiply the products of society a hundredfold."

transformation in human history in the primitive accumulation. However, while Polanyi emphasizes the "commodification of human and nature" in this historical process, I focus more on the dissolution of the community, which is a unique mode of production.

However, it may be possible to emphasize the pastoral aspect in this way only in Western Europe, where the primitive accumulation developed spontaneously. Luxemburg (1913) is based on the British invasion of India, French colonial policy in Algeria, and the Opium War. Her book states that under colonization by Western powers, the self-sufficient natural economy protected by the community was violently dismantled. It also describes in detail the process by which the subsequent introduction of the commodity economy turned the former self-sufficient natural economy into a commodity purchaser for capital and a seller of raw materials and labor. In other words, because the primitive accumulation in the colonies was carried out in the absence of the intrinsic development of the productive forces, and because the primitive accumulation was carried out "from above" without waiting for the intrinsic development of the productive forces, the violent process came to the fore. There, rigid social relations in the traditional forms of property keep the labor-power in such a tight fetter that, it takes several decades for a proletariat that is subject to capital. (Luxemburg 1921)

4. Conclusion—The "Beginning" and "End" of Primitive Accumulation

In short, the concept of primitive accumulation includes the dismantling of the community and of its "principle of survival" as an important element. As long as the community is the foundation of a pre-capitalist societies, it is related to the "beginning of primitive accumulation." [19] Furthermore, even if we simply say "the beginning of the primitive accumulation," it is not something that can be strictly determined as a certain point in time, as is clear from the descriptions so far. Rather, it should be expressed as a "starting period" with a certain period width. In this sense, the "starting period" of the primitive accumulation has itself a "beginning" and an "end." As mentioned in Section 2, the "principle of survival" of the community has various aspects, including law and religion. However, since the symbol of the "principle of survival" is the common land, the "beginning" of the "starting period" of primitive accumulation is when the usurpation of the common land begins. The "end" is when the usurpation is completed. In England, in the 19th century, "the very memory of the connection between the agricultural laborer and the communal property had, of course, vanished." (Marx 1867) It can be said that General Enclosure Acts marked the end of the "starting period." In this sense, the "starting period" of British primitive accumulation was a long-term process spanning more than three centuries from the end of the fifteenth century. Along with the denial of community-based modes of production, or as a premise, "divorcing the producer from the means of production" progressed. Moreover, at some point after this separation and capital accumulation has progressed, we should not be concerned with the end of the "beginning period" of primitive accumulation, but now with the "end" of primitive accumulation itself. And only when we recognize the "beginning" and "end" of the primitive accumulation, we will be able to understand the connotation of the concept of primitive accumulation as a whole.

What, then, is the "end" of primitive accumulation? In the previous chapter,

19 "The first aim of capitalism is to isolate the producer, to sever the community ties which protect him, and the next task is to take the means of production away from the small manufacturer." (Luxemburg 1913)

while taking the primitive accumulation theory as a precondition for the theory of capital accumulation, I argued that when the creation of the labor force population through the increase of the organic composition of capital and the business cycle became the main supply route for the labor force, primitive accumulation is said to have ended. This means that at this time the capitalist society acquired its own labor procurement mechanism and was thereby established. The direct producers protected by the community of "principle of survival" were separated from the land, and the working population that had been staying in rural areas was washed away to the cities. As a result, it is no longer possible to compensate for the increased demand for labor in cities with the supply of labor from farmers. The rise in the wage level resulting from this will induce the development and introduction of labor-saving machines, which will lead to the increase of the organic composition of capital.[20] "The limit to using a machine is fixed by the difference between the value of the machine and the value of the labor-power replaced by it." (Marx 1867) Therefore, the

---

20 According to Fujise (1985), "the starting point of the primitive accumulation = the beginning of the dissolution of feudal society." Regarding the end point of the national dimension, he said, "Even in the developed capitalist countries, except for England, land grabbing from farmers is not carried out thoroughly, and various types of farmer management = land ownership continue to exist widely. However, this kind of farmer management = land ownership has been transformed into a form suitable for the accumulation of capitalism, and is governed by its rhythm. When talking about the establishment of capitalism in a single country, it would be presumed that such conditions were established systematically. In this case, it can be said that the primitive accumulation has come to an end in that country." Regarding the termination at the global level, he said, "The beginning of the twentieth century, when most of the world's regions were integrated and reorganized into the world market and the capitalist world system was established, can be regarded as the final point." I have learned a great deal from his raising the issue of putting the starting and ending points of the primitive accumulation the table for discussion. On the other hand, his approach to starting and ending points is not based on the arrangement of the primitive accumulation concept, so unfortunately it is difficult to say that his approach is clear because it reflects the ambiguity inherent in this concept.

higher the wage level, the more machines are employed. Therefore, in order for the increase of the organic composition of capital to become the main route for supplying labor, it is necessary that the supply of labor from peasants disappears and that the special farming character of labor wages disappears. In other words, a rise in labor wages to encourage the adoption of machines is an essential condition.

Here, let us look back at the diversity of forms of primitive accumulation and historical periods across countries. In this respect, the primitive accumulation, which is the original process of separation between land and labor, which forms the historical starting point of the relationship between capital and wage labor, has traditionally been regarded as a social phenomenon belonging to the period from the end of feudalism to the period of mercantilism. However, as touched upon at the beginning of this chapter, Marx's writings also contain cautionary notes regarding (1) the diversity of primitive accumulation by country and (2) the regional scope of application of the description of primitive accumulation in "Capital." They also contain descriptions suggesting the possibility of the existence of regions outside the above period. "The expropriation of the agricultural producer, of the peasant, from the soil, is the basis of the whole process. The history of this expropriation, in different countries, assumes different aspects, and runs through its various phases in different orders of succession, and at different periods." (Marx 1867) According to "Capital," the era to which primitive accumulation belongs differs from country to country. Marx also stated that "in England alone, which we take as our example, has it (=primitive accumulation) the classic form." (ibid.) Furthermore, in the French version of "Capital," which is not used in the current version of Engels's "Capital," but which Marx himself made during his lifetime, he wrote, "*Mais tous les autres pays de l'Europe occidentale parcourent le même movement,* (However, all the rest of

Western Europe other countries go through the same movement)." (Marx 1872-75) In his famous letter to Zasulich in 1881, while citing the description of the French version, he clearly limited the scope of application of the British primitive accumulation in "Capital" to Western Europe. Furthermore, the reason for this limitation is stated as follows.

"In this movement of the West, the transformation of one form of private property into another is at stake. Contrary to this, Russian peasants want to transform communal ownership into private ownership." (Marx 1881)

The establishment of capitalist society in Western Europe was achieved by transforming private ownership based on self-labor as a "necessary transition point" into capitalist private ownership. However, Russia is trying to transform communal ownership directly into capitalist private ownership. Therefore, in the latter, it is said that we must assume a different type of primitive accumulation from that in Western Europe. What Marx is directly addressing here is the typological difference between Western European and Russian primitive accumulations. More generally, however, it is in line with Marx's intent to think that a typology of national primitive accumulations can be envisioned. However, the more we recognize the existence of such diversity, the more we need to clarify the general concept of primitive accumulation itself. So far in this chapter, it is referred to as (1) the destruction of the community of the "principle of survival" and (2) the process of forming the preconditions for accumulation in an established capitalist society, namely the preconditions for the functioning of labor supply through the creation of surplus-population.[21]

Going one step further, how should we understand these two milestones in

21 Tamagaki (1969) argues that within Marx's theory of accumulation itself, the perspective of decomposition of pre-capitalist elements overlaps with the perspective of creating a relative surplus-population through the increase of the organic composition of capital.

a unified way? In other words, what exactly is the primitive accumulation whose beginning and end are dictated by these two merkmals? When thinking about this, it is important to understand that the "end point" of primitive accumulation is also the point at which the tyrannical control of capital over labor in the production process, i.e., the "real subsumption of labor under capital," is established. First, the moment for capital's tyrannical control over labor in the production process is brought about by the introduction of machinery itself. This is because manual skill is dismantled, labor is simplified, and the spiritual element of the labor process is alienated from labor and transformed into the power of capital. Capital's dominance over labor in the process of production arises from the deepening of labor's dependence on the means of labor, the overturning of subject and object, and the alienation and deprivation of the mental elements of labor. This dominance is further enhanced by the relative surplus-population due to the business cycle and to the increase of the organic composition of capital, which leads to the constancy of competition among workers and the expansion and deepening of the social dependence of labor on capital. In this way, the "end point" of primitive accumulation is also the point at which the tyrannical rule of capital, i.e., the "real subsumption of labor under capital," is established. If so, it is also the time when the dismantling of society based on the "principle of survival" is finally completed. Indeed, the "principle of survival" itself is lost along with the community. However, the "social principle" [22] based on the social solidarity of people, which remained after that, was further damaged by the activation of the law of relative surplus-population and the constant competition among workers. Furthermore, around that time, the non-capitalist elements centered

22 Yamada (2008) provides an understanding of "social principle." Also, Baba's (1981) contrasting understanding of capitalist principle and "social" principle is helpful.

on peasant agriculture that had previously functioned as social buffer disappeared, and this itself constituted the loss of the "social principle." If we understand the meaning of "end point" in this way, we can understand the primitive accumulation itself, that is, the whole from the beginning to the end, as a process in which the "principle of survival" is denied in the first place, and furthermore, it is finally buried. I would like to conclude this chapter by presenting such an understanding of the concept of primitive accumulation.

Chapter 3 : On the Controversy over the Concept of Primitive Accumulation

1. Theme

Chapter 24, "The so-called primitive accumulation," in the current version of "Capital," is a chapter with a complex structure that interweaves theoretical provisions with a variety of depictions of the actual situation, mainly based on British history.

In the first section, the concept of primitive accumulation is defined as follows. "The so-called primitive accumulation, therefore, is nothing else than the historical process of divorcing the producer from the means of production." (Marx 1867) "The expropriation of the agricultural producer, of the peasant, from the soil, is the basis of the whole process." (ibid) Section 2 describes the actual situation of land grabbing from farmers. Section 3 presents the measures taken to cultivate the newly created workers to work on terms favorable to capital. Sections 4, 5, and 6 discuss the creation of capitalists in agriculture and industry and the problems of market formation for capitalist industry. The final section, Section 7, is a macroscopic historical dynamic theory of capitalist society starting with primitive accumulation. In addition, in Section 7, the definition of the primitive accumulation concept in Section 1 is reproduced at the beginning as follows. This shows that this definition is consistently maintained throughout Chapter 24. "What does the primitive accumulation of capital, i.e., its historical genesis, resolve itself into? In so far as it is not immediate transformation of slaves and serfs into wage-laborers, and therefore a mere change of form, it only means the expropriation of the immediate producers, i.e., the dissolution of private property based on the labor of its owner." (ibid.)

Chapter 24 contains descriptions of these diverse historical processes. However, in that chapter, primitive accumulation in its own sense is the process of separation between producers and the means of production, which is the historical premise for the conversion of money into capital. The actual state of this process is described in Section 2. However, Marx emphasized the significance of the violence that was used in the process of separating farmers from the land, and it is true that violence had an important meaning in the primitive accumulation. On the other hand, however, there are also historical periods in which the differentiation and disintegration of peasant strata, determined by competition in productivity among farmers and the attraction of labor by non-agricultural capital, has an important meaning in this process of separation (see previous chapter of this book). Since it includes such "peaceful" process, the concept of primitive accumulation also encompasses agricultural structural dynamics and agricultural questions.

On the other hand, the descriptions from Section 3 onwards include the cultivation of workers, the formation of the domestic market, the creation of industrial capitalists, and the policies of mercantilism. They are, after all, depictions of the historical process by which British capitalist society was born. Although primitive accumulation is a process of separation between producers and the means of production, since it is a historical process, various things occur along with it in society. "Capital" captures primitive accumulation within this expansion. However, what must be noted here is that the depiction of the primitive accumulation itself and the description of the primitive accumulation period are different. As we will see later, "Capital" distinguishes the elements included in the latter from the primitive accumulation itself, calling them "elements (moments) of primitive accumulation." [23] Furthermore, the primitive accumulation period in British history largely overlaps with the manufacturing

23 "Moments" are dynamic factors that cause things to change and develop.

period, which is said to be "the period from the middle of the 16th century to the last third of the 18th century." Therefore, in Chapter 24, although the term "primitive accumulation period" as distinct from the manufacturing period does not appear, the word "manufacture" appears frequently. Additionally, the phrase "manufacturing period" appears a total of six times in Sections 3, 5, and 6. However, the historical period in which "primitive accumulation" is located can vary from country to country, therefore in extending this concept to other countries, I want to give citizenship rights to the more general term "primitive accumulation period." Also, needless to say, every "period" has a beginning and an end. A "period" is an era defined by its beginning and end. Therefore, by arriving at the concept of the primitive accumulation period, a further question that must be asked is what constitutes the beginning and end of the primitive accumulation period. Regarding this point, please refer to the previous chapter of this book. In any case, through Chapter 24, we will be able to recognize the internal structure of the concept of primitive accumulation while distinguishing these three concepts; primitive accumulation (process),[24] primitive accumulation period, and various elements for primitive accumulation. This is a crucial point in order to understand the concept of primitive accumulation (process) as a consistent thing (**Table 3-1**).[25]

---

24 In this chapter, the two terms "primitive accumulation" and "primitive accumulation process" are used interchangeably.

25 Iwasaki (2015) argues as follows in his commentary on Yamazaki (2014b). "It goes without saying that primitive accumulation is the process of creating capital relations, and in particular, the process of creating wage labor is the primary one. However, today's violent destruction of nature and humans by states and international organizations (due to neoliberal policies) and the resulting 〈creation of new capital relations〉 on a global scale can be said to be a 〈primitive accumulation function〉. If that were the case, the concept of primitive accumulation would be expanded even further." In this text, a new concept of "primitive accumulation function" appears. When using this term, Professor Iwasaki may be thinking of the same content as my "primitive accumulation period," which consists of "primitive accumulation + various

Table 3 -1 Primitive accumulation period

Primitive accumulation period	Primitive accumulation —Separation process between direct producers and means of production (Fracture of community, Decomposition of small commodity producers) —Chapter 24, Section 2 Elements of primitive accumulation —① Invocation of state power to control wages and extend the working day (essential moment) —Chapter 24, Section 3 ② Capital accumulation (main moment) and the formation of industrial capitalists—Chapter 24, Sections 4 and 6 ③ Formation of domestic market for industrial capitalists—Chapter 24 Section 5

Note 1) Chapters and sections correspond to the current volume 1 of "Capital".

In Chapters 1 and 2 of this book, I defined the concept of primitive accumulation as a process of separation between producers and means of production as follows, keeping in mind the relationship between this concept and the description in Chapter 23 of "Capital." Primitive accumulation begins with the destruction of the community based on the "principle of survival." Moreover, it is a process in which the precondition for inherent accumulation in an established capitalist society is formed. This is a prerequisite for the functioning of labor supply through the raise of the organic composition of capital and the creation of a relative surplus-population through business cycles.

By the way, the following criticism can be expected against this understanding of the concept of primitive accumulation. "Primitive accumulation must have the

---

elements of primitive accumulation." The professor calls this a "further expansion of the concept of primitive accumulation," but I regard the various elements of primitive accumulation and the period of primitive accumulation as different concepts from primitive accumulation itself. I would like the professor to further explain the relationship between "primitive accumulation" and "primitive accumulation function."

aspect of creating a productive working class, and at the same time the aspect of accumulating monetary wealth that can be converted into capital. This aspect is neglected as the concept definition." And this is not just a hypothetical story. A certain scholar actually stated this in a discussion with me. Now, let's call this academic person Mr. X.

At first glance, it seems to me that Mr. X's argument is unreasonable. First of all, "the creation of a productive working class" and "the accumulation of monetary wealth" [26] can be conceptually and practically two different things. These two things coincide in the special case where capitalists and workers are formed through the bipolar decomposition of direct producers, and capitalists become capitalists using their own funds that they have accumulated. These things are inconsistent with the historical picture of the emergence of capitalist society in "Capital," which we will see later. In fact, they do not seem to match the facts of British economic history. [27]

---

26 "Monetary wealth" is a form of "wealth" that is the opposite of "land assets," which was the typical form of "wealth" in societies from pre-capitalism to the primitive accumulation period. It is money capital that has been accumulated through pre-capitalist commerce and pre-capitalist usury and is being transformed into industrial capital. That is, in two capital circulation formulas, G−W−G'・G−W−P−W'−G' or G−G'・G−W−P−W'−G', "monetary wealth" is G' that is in the middle of each formula. Furthermore, G−W−G' and G−G', which begin these capital circulation formulas, indicate the movement of capital in pre-capitalist society. Furthermore, this term ("monetary wealth") is at odds with mere "money" as used by Rosenberg (discussed later), as it implies that it is the result of capital accumulation.

27 These seem similar to the historical picture of the emergence of capitalist society drawn by Otsuka (1956a), but they are different. Whereas Otsuka is concerned with the central figures in the formation of a capitalist society, what is at issue here is the "accumulation of monetary wealth." Regarding the difference between the two, Dobb (1946) says the following. 〈Capital to finance new technology came primarily from merchant houses and commercial centers such as Liverpool. However, many of the people who directed the new factory-based industry and took the lead in its expansion were people of lowly background. They came from the former craft master and yeoman class. They also had some capital accumulated, which they increased by forming alliances

How on earth is it possible to sum up these two things, which are distinct from each other, into a single concept called primitive accumulation? In fact, "Capital" also defines the concept of primitive accumulation in the manner described above. It does not say that primitive accumulation consists of two elements: "creation of a productive working class" and "accumulation of monetary wealth." Instead, this classic says primitive accumulation is "the creation of a productive working class." It can be read as "primitive accumulation = a historical separation process between producers and the means of production," and "primitive accumulation ≠ other social phenomena."

Furthermore, although the problem of "the creation of industrial capitalists" being discussed in Sections 4 and 6 seems similar to the problem of "accumulation of monetary wealth," these are mutually different problems. The problem of "the accumulation of monetary wealth" is the question of how "monetary wealth" that did not previously exist was accumulated. On the other hand, the problem of the "creation of industrial capitalists" is how the already existing "monetary wealth" is combined with idle workers and transformed into industrial capital.

By the way, if Mr. X's views are unique to him, there may be no need to bring them up here. However, as we will see later, Mr. X's view actually seems to have its roots in the confusing descriptions in the authoritative commentary on "Capital." Therefore, I cannot deny the possibility that his view still has a certain degree of social influence. Therefore, the purpose of this chapter is to critically examine X's view, which attempts to understand the concept of primitive accumulation by incorporating the element of "accumulation of monetary wealth." Furthermore, the task is to clarify how this contamination

---

with wealthier merchants. This is a well-known fact.〉 In short, Dobb argues that wealthy farmers and artisans became industrial capitalists, while receiving the necessary capital from large merchants.

occurred. It goes without saying that when using the concept of primitive accumulation, especially in empirical science such as agricultural economics, it is necessary to accurately capture its content.[28]

The subject of this chapter is the current "Capital." As is well known, in addition to the current version, there are various versions of Volume 1 of "Capital" that were modified by Marx and Engels. In particular, the French version is also called "the last Capital" because it is the first volume that Marx edited last. There, some particularly significant changes have been made to the section that corresponds to the current version of Chapter 24 in question. However, many of these changes are not reflected in the current version. Therefore, although researching the theory of primitive accumulation in the French version of "Capital" could be an issue in its own right, this chapter does not go that far.

2. Rosenberg's "Commentary on Capital" and Kawakami's "Introduction to Capital"

The first thing I will examine here is Rosenberg's "Commentary on Capital" (1931-1933). The book has long been used as a commentary on "Capital," and has had a great influence on the understanding of this classic in Japan.

28 Ito (1982) states: "Marx's primitive accumulation of capital provided the basic premise for the establishment of capitalism. The fundamental of this concept was to dismantle peasants' feudal land tenure rights and cultivation rights, to expropriate land, often through violence and political power, and to form modern land ownership." Although this explanation may be sufficient to give an image of the concept of primitive accumulation, it is unsatisfactory when it comes to using this concept in empirical analysis. There is no expression such as the "fundamental" of primitive accumulation in the current "Capital," so what is the meaning of "fundamental?" What are the elements other than the "fundamental?" Also, what is the relationship between "fundamental" and elements other than "fundamental?" Could these points be a problem?

Therefore, it is worth considering whether Mr. X's earlier view was influenced in some way by this commentary. In fact, in the "Commentary on Capital," it appears that the element that is currently in question, "accumulation of monetary wealth," has entered into the concept of primitive accumulation. Therefore, let's begin our examination of the book by confirming this contamination. In the commentary on Chapter 24, "the so-called primitive accumulation" of the book, the author states the following about the research object of this chapter, and claims that it deals with two problems.

"Chapter 4 ('Capital Volume 1' —citer) argues that money becomes capital because a portion of it is converted into labor-power that is separated from the means of production and can be sold by (legally) free workers. Already then the question arose as to where such a labor-power would come from."

Here, I will address the issue related to this kind of "separation of producers and means of production" as the first question.

"Furthermore, as research progressed, it became clear that not any amount of money could be converted into capital. In other words, the minimum amount that could be advanced for production was shown to be far greater than the medieval maximum, that is, the maximum amount of money that medieval guild masters needed to carry out their work. Therefore, the next problem occurred. How and by what means did individuals accumulate an amount of money in their hands that exceeded the 'medieval maximum?'"

I will address the issue related to this kind of "accumulation of monetary wealth" as the second question.

The idea in the "Commentary on Capital" that the theory of primitive accumulation deals with these two issues seems to be reflected in Mr. X's earlier understanding of the concept of primitive accumulation.

Furthermore, the "Commentary on Capital" states that when these two problems are combined into one, the following problem is obtained.

"How did it happen that the means of production were separated from some people and appropriated by others, so that the former became proletariats and the latter capitalists?"

Let's now take this as a general problem.

If I were to make my own comments on the reasoning in the "Commentary on Capital," I would firstly say that when the previous two problems are combined, it is difficult to understand why such a general problem arises. This is because the second question concerns the "accumulation of monetary wealth" by people who become capitalists, whereas in the general problem, what people who become capitalists gain is what direct producers have lost. It is a means of production. Between the two issues, the targets which people who would become capitalists gather have changed. Since land is the main thing that direct producers lose, land is the main thing that people who would become capitalists accumulate in the general problem. Since the "accumulation of monetary wealth" and the accumulation of land are different matters, I do not think that the general question can be derived directly from the second question.

In any case, if we approach the subject of Chapter 24 as if it were the "Commentary on Capital," then this means imposing a strange role on "Capital." This is because, as mentioned earlier, the definition of the concept of primitive accumulation in "Capital" can be said to correspond to the first question, but it is not consistent with the second question. Certainly, in Chapter 24, Section 6, there is a vivid description of the commercial warfare of European countries on the globe and the progress of capital accumulation through public bonds during the primitive accumulation period (manufacturing period). However, it is not a description of the first accumulation of monetary wealth, exceeding the "medieval maximum," which is being transformed into industrial capital. Rather, it can be thought of as a depiction of the historical background of primitive accumulation as a historical process. In short, if the subject of Chapter 24 is set as in "Commentary on Capital," "Capital" is not consistent with the subject in terms of the definition of the concept of primitive accumulation and the content described in that chapter.

What I have even more doubts about is the origin of the second question. How did the "Commentary on Capital" derive the second question? It is clearly stated that the first question was derived from Chapter 4 of "Capital." However, regarding the second question, the "Commentary on Capital" only states that "as research progressed," making the origin of the question ambiguous. I have no choice but to think that it is somewhere between Chapters 4 and 23, but I think it is probably the following statement within Chapter 9, "rate and mass of surplus-value."

"From the treatment of the production of surplus-value, so far, it follows that not every sum of money, or of value, is at pleasure transformable into capital. To effect this transformation, in fact, a certain minimum of money or of exchange-value must be pre-supposed in the hands of the individual possessor of money or commodities." (Omission) "The guilds of the middle ages therefore tried to prevent by force the transformation of the master of a trade into a capitalist, by limiting the number of laborers that could be employed by one master within a very small maximum. The possessor of money or commodities actually turns into a capitalist in such cases only where the minimum sum advanced for production greatly exceeds the maximum of the middle ages." (Marx 1867)

As can be seen, "Capital" states here that the medieval maximum existed not because the masters had only a small amount of money in their hands. Rather, the maximum existed because there were the guilds that forcibly limited the maximum that could be advanced for production. This system artificially prevented the use of accumulated "monetary wealth" in production.

Therefore, the situation may have been the opposite of what the "Commentary on Capital" assumed. "Monetary wealth" has been borne by commercial peoples living between communities since ancient times, and may have already accumulated to a considerable extent in the middle ages. However, this wealth may have been blocked by the guilds and prevented from being invested in production. And, as we will see later, "Capital" appears to rather stand on this position.

We have seen above that the two questions that "Commentary on Capital" claims to be the subject of Chapter 24 and their integration contain difficulties that cannot be easily accepted. Difficulties were related to consistency with the content of "Capital" and the process of deriving questions.

Next, let's look at Hajime Kawakami's "Introduction to Capital" (1929), which was published at the same time as "Commentary on Capital" and was a pioneer in Japanese commentary on "Capital." Does the book say something different from the "Commentary on Capital?" In relation to Chapter 24, "Introduction to Capital" states the following: "A further question for us is how the starting point, the first capitalists or wage workers, came into existence."

Kawakami also seems to see the same issue as Rosenberg in Chapter 24, such as "the emergence of the first capitalists and workers." However, regarding the creation of the "first capitalists," Kawakami does not set such an in-depth condition as Rosenberg did, such as "the accumulation of monetary wealth that exceeds the medieval maximum." Kawakami simply states, "what we are now concerned with is the process by which capitalist production can reach such a degree of independence." However, unlike Rosenberg, Kawakami does not discuss in depth the conditions for "capitalist production can reach an independence."

Furthermore, what Mr. X's discussion earlier focused on was the definition of the concept of primitive accumulation. On the other hand, the issue addressed in this section of this chapter is the theme in Chapter 24 of "Capital." These two are different things. In fact, even the "Commentary on Capital" respects Marx's definition of the concept of primitive accumulation itself. It correctly states: "Primitive accumulation is the expropriation of the means of production from producers with the aim of converting them into wage workers." However, as mentioned earlier, this correct understanding is inconsistent with the theme setting for Chapter 24 of "Capital" in this commentary. Rosenberg has two conflicting souls between the setting of the subject in Chapter 24 of "Capital" and the definition of the concept of primitive accumulation. Furthermore, this contradiction in understanding regarding Chapter 24 in authoritative commentary is reflected in the way people like Mr. X still view the concept of primitive accumulation while respecting Rosenberg's agenda setting. How did this contradiction in recognition develop dynamically in the scholars after Rosenberg?

3. Commentaries from the 1960s Onwards

As we saw above, in the "Commentary on Capital," published around 1930, Rosenberg identified the two issues of Chapter 24 as "the accumulation of monetary wealth and the creation of workers." In this section, we will look at how this understanding changed in the commentary books subsequently published in Japan. However, it should be noted in advance that this is not an exhaustive review of all of the numerous commentaries on "Capital." I will limit myself to picking up works from the 1960s, 70s, and 80s that I think are representative of each era. [29]

29 Naturally, I cannot escape criticism that my selection of commentaries is arbitrary. In addition, the text jumps from around 1930 to the 1960s. Therefore, although it is not a commentary book, I would like to quote here the definition of primitive accumulation in the "Textbook of Economics" by the Institute of Economics of the Academy of Sciences of the Soviet Union (1954), which is

First, let's look at "Research on Capital" (1967), edited by Kozo Uno. It states the following about the issues in Chapter 24. [30]

"This chapter describes the process of primitive accumulation through the history of England, where it followed a typical process. It becomes clear that this process is as follows. On the one hand, it is the process of converting society's means of living and production into capital, and on the other hand, the process of converting direct producers into wage workers. Alternatively, it is the process of historical separation of workers from the means of production."

First of all, it should be noted that the issue of "accumulation of monetary wealth," which Rosenberg had as an issue for the primitive accumulation theory, is not raised here. On the one hand, there is a phrase that says, "the process of converting society's means of living and production into capital," and on the other hand, there is also a description of "the process of historical separation of workers from the means of production." The theme of Chapter 24 is the two elements of capital formation and worker creation.

Next, let's look at "Learning about Capital" (1977), edited by Kinzaburo Sato et al. There is no place where it explicitly states, "this is the problem of primitive accumulation theory." However, the following description exists regarding the conceptual definition of primitive accumulation.[31]

"The primitive accumulation of capital is essentially nothing but the fundamental creation of capital relations. Certainly, it must have the aspect of a

---

located between these periods. "A mass of proletarians, was created and wealth was accumulated in the hands of a few because the means of production were forcibly taken away from small producers. The process by which the means of production, land, and tools of production were separated from the producers was accompanied by countless acts of plunder and cruelty. This process is called the primitive accumulation of capital because it occurred before the creation of capitalist big business."

30 The person in charge of writing this part is Tsuyoshi Sakurai.

31 The person in charge of writing this part is Seiji Mochizuki.

preliminary accumulation of a certain amount of capital to be invested in industry in the future. But even more than that, there must be the decisive factor that makes the holder of this money an industrial capitalist, that is, the process of creating a horde of wage workers. Bourgeois economists believe that once the means of production or capital is accumulated, the workers who turn it into capital will come from somewhere."

Here, the perspective of "accumulation of monetary wealth" is emphasized by bourgeois economists. While "Capital" criticizes such a viewpoint, it emphasizes the significance of "the creation of workers" in the historical formation of capital relations. In other words, the focus on the "creation of workers" in "Capital" comes from the perspective of criticism of economics.

Finally, let's look at "System of Capital" (1985), edited by Ryozo Tomizuka et al. The issues in Chapter 24 are stated as follows. [32]

"It has already been made clear in Volume 1, Part 2 of 'Capital,' 'the transformation of money into capital' that capital is something that is created by converting money that has been accumulated over a certain amount. However, it has become clear that this transformation requires the condition that 'money holders encounter free workers in the commodity market.' However, the necessity of this condition was merely clarified, and the elucidation of the process by which this condition was established was left as a subject for later research.

This issue is now being addressed head-on."

As can be seen, the text echoing Rosenberg's "medieval maximum" is reproduced here: "Capital is something that is created by converting money that has been accumulated over a certain amount." However, what is even clearer is that this "accumulation of money" is no longer listed as an issue in Chapter 24. Here, the task of Chapter 24 is to elucidate the process by which the necessary conditions for money to be converted into capital are established,

---

32 The person in charge of writing this part is Naomichi Hayashi.

that is, "money holders encounter free workers in the commodity market."

What has become clear from the previous section to this section can be summarized as follows. First of all, it was the "Commentary on Capital" (1931-1933) that clearly established the "accumulation of monetary wealth" as well as the "separation of producers and the means of production" as the subject of Chapter 24. Second, however, "Research on Capital" (1967) already emphasized the "separation of producers and means of production," while the issue of "accumulation of monetary wealth" receded into the background. Third, in "Learning about Capital" (1977), it was stated that this sharp emphasis stemmed from the perspective of Marx's criticism of economics. Finally, in "System of Capital" (1985), the subject of Chapter 24 has been narrowed down to "separation of producers and means of production." Not only has the issue of "accumulation of monetary wealth" been dropped from the discussion, but even the formation of capital relations, which had been the focus of "Research on Capital," is no longer raised as an issue. In this series of discussions, Rosenberg's "accumulation of monetary wealth" was dropped from the topic of Chapter 24 at an early stage. At the same time, the issue was gradually narrowed down to "separation of producers and means of production." If this way of looking at the flow of the discussion is valid, then the convergence to regarding the issue in Chapter 24 as "separation of producers and means of production" was not achieved through critical examination of previous research. This is a problem. This seems to be the reason why view like Mr. X's, which includes the "accumulation of monetary wealth" in the concept of primitive accumulation, has persisted forever.

4. Economics system and Criticism of Economics

By the way, if we were to understand the concept of primitive accumulation as described above, the question becomes how to position the various other descriptive contents written in Chapter 24 in relation to this conceptual definition. The problem I am aware of when undertaking this kind of work is whether the roots of Mr. X's views can be found in the contents of these descriptions.[33] To state the conclusion in general terms, Chapter 24, Section 2 describes the primitive accumulation process itself. In addition, in Sections 3 to 6, the background of primitive accumulation as a historical process is described as various elements of primitive accumulation. But we will look at these in more detail later.

Before looking at these contents, I would like to raise an issue in this section in advance that Chapter 24 does not only develop the concept of primitive accumulation. In addition, there are also passages that criticize the economics of property, and this aspect is described with as much importance as the development of the concept of primitive accumulation. Of course these two aspects are closely related, therefore they are discussed in the same chapter. However, it is also true that this fact makes it difficult to read this chapter in accordance with the purpose of this classic as a system of economics (elucidation of economic laws of movement in capitalist society), which "Capital" states in its preface to Volume 1.

In particular, at the beginning of Chapter 24, the problem is raised that "the

33 In "Capital," logic develops upward, from the abstract to the more concrete. There, in some cases, things that appear to be contradictory on the surface to previous statements are later stated. This provides Marx's critics such as Böhm-Bawerk (1896), who do not understand the methodology of this classic, a convenient foothold to advance their arguments. What Sections 4 and 5 of this chapter examine is the relationship between the concept of primitive accumulation mentioned in Chapter 24, Section 1 of Volume 1 of "Capital" and what is said elsewhere in the same chapter. The question is whether there are any contradictions between them. Furthermore, if a contradiction exists, does that not provide a basis for Mr. X's argument? What this chapter extracts from "Capital" after such consideration is the distinction between these three concepts; the period of primitive accumulation, the primitive accumulation, and the elements of primitive accumulation.

whole movement, therefore, seems to turn in a vicious circle, out of which we can only get by supposing a primitive accumulation (previous accumulation of Adam Smith) preceding capitalistic accumulation; an accumulation not the result of the capitalist mode of production but its starting-point," and the following conclusion is also stated: "In actual history it is notorious that conquest, enslavement, robbery, murder, briefly force, play the great part." Perhaps the existence of such a description in "Capital" is one of the reasons behind Mr. X's point that "primitive accumulation must have both the aspect of creating a productive working class and the accumulation of monetary wealth that can be converted into capital." However, if you read "Capital" carefully, you will find that it is not clearly stated that the above "accumulation" is "the accumulation of monetary wealth." Therefore, there is room for interpretation as to what constitutes "accumulation," and based on the above definition of primitive accumulation, the content of "accumulation" is rather the accumulation of land that corresponds to the land taken from farmers. In fact, the pre-capitalist "wealth" accumulated by the "élite" in the "legend of theological original sin," as described in the following sentence in the second paragraph from the beginning of Chapter 24, is not "monetary wealth." It's land.

"This primitive accumulation plays in Political Economy about the same part as original sin in theology. Adam bit the apple, and thereupon sin fell on the human race. Its origin is supposed to be explained when it is told as an anecdote of the past. In times long gone by there were two sorts of people; one, the diligent, intelligent, and, above all, frugal élite; the other, lazy rascals, spending their substance, and more, in riotous living, (Omission) Thus it came to pass that the former sort accumulated wealth, and the latter sort had at last nothing to sell except their own skins."

However, taking a step back, I think that even if "accumulation of monetary wealth" is important, it should be viewed as something mediated by land accumulation.

In fact, Chapter 24, Section 2, after giving a detailed account of land expropriation from farmers, concludes with the following line, suggesting that land accumulation has become a lever for capital accumulation.

"The spoliation of the church's property, the fraudulent alienation of the State domains, the robbery of the common lands, the usurpation of feudal and clan property, and its transformation into modern private property under circumstances of reckless terrorism, were just so many idyllic methods of primitive accumulation. They conquered the field for capitalistic agriculture, made the soil part and parcel of capital, and created for the town industries the necessary supply of a 'free' and outlawed proletariat." (Underlining is by the quoter.)

However, this way of reading Chapter 24, which emphasizes land accumulation, does not seem to be my own original reading. Dobb (1946) emphasizes the land accumulation aspect of primitive accumulation, and arguing that the "accumulation of monetary wealth" was important only insofar as it contributed to land accumulation.

By the way, even if we understand the concept of primitive accumulation with emphasis on the "separation of producers and means of production," the following problem remains. How was the initial "monetary wealth" that was transformed into industrial capital, the starting point of capitalist society, formed? The answer provided by "Capital" to this question is probably that "such monetary wealth already existed before the period of primitive accumulation." "The pre-Flood form of capital" was created from time immemorial by commercial peoples whose main arena of activity was between communities, "like the gods of Epicurus," "or rather like the Jews in the pores of Polish society." When the time was ripe, when the separation between producers and the means of production spread socially, and when the investment restrictions of feudal society were removed, the process by which the capital took control of the production process was the creation of capitalist society. This was one of Dobb's main points in the transition debate, which revolved around the Dobb-Sweezy debate.[34] In fact, in Chapter 24, Section 6, "Genesis of the Industrial Capitalism," there is the following statement that seems to support this view.

"The money capital formed by means of usury and commerce was prevented from turning into industrial capital, in the country by the feudal constitution, in the towns by the guild organization. These fetters vanished with the dissolution of feudal society, with the expropriation and partial eviction of the country population. The new manufactures were established at sea-ports, or at inland points beyond the control of the old municipalities and their guilds."

As we have seen in this section, Chapter 24 contains two aspects of "Capital," one as an economics system and the other as a critique of economics, making it difficult to distinguish between them. However, as we will see below, it is difficult to distinguish between the primitive accumulation itself and the depiction of the primitive accumulation period.

5. Usages of the Term of "Primitive Accumulation" in Chapter 24, Sections 3-6

In Chapter 24, Sections 3 to 7, there are several descriptions that include the term "primitive accumulation." Let us take a look at them here. However, since the word "primitive accumulation" does not appear in Section 4, we will

34 This controversy originated with Dobb (1946). For the subsequent development of the argument, see Dobb et al. (1976). See also note 27 in this chapter.

focus on Sections 3, 5, and 6 here. The last section, Section 7, has already been mentioned.

Section 3 contains the following statement: "The bourgeoisie, at its rise, wants and uses the power of the state to 'regulate' wages, i.e., to force them within the limits suitable for surplus-value making, to lengthen the working-day and keep the laborer himself in the normal degree of dependence. This is an essential element of the so-called primitive accumulation." In the period of primitive accumulation, the capitalist law of population has not yet begun to operate. Under such circumstances, it is necessary to exercise state power to suppress wages and extend the working-day. "Capital" describes this situation as "an essential element of the so-called primitive accumulation."

Section 5 begins as follows: "The expropriation and expulsion of the agricultural population, intermittent but renewed again and again, supplied, as we saw, the town industries with a mass of proletarians entirely unconnected with the corporate guilds and unfettered by them; a fortunate circumstance that makes old A. Anderson (not to be confounded with James Anderson) in his 'History of Commerce,' believe in the direct intervention of Providence. We must still pause a moment on this element of primitive accumulation." The transformation of direct producers into workers during the primitive accumulation period is also a process in which a domestic market is formed as their means of living and labor are transformed into the material elements of capital. Needless to say, this was the point that Lenin (1893) emphasized in his dispute with the Narodniki, who argued that the downfall of the peasantry would lead to a contraction of the market.

By the way, in the text of Section 6, the word "primitive accumulation" appears seven times, which is the most for any section in Chapter 24. Incidentally, it appears a total of 10 times in other sections. The breakdown is 5 times in the 1st section, 1 time in the 2nd section, 1 time in the 3rd section, 0 times in the 4th section, 1 time in the 5th section, and 2 times in the 7th section. Furthermore, as we will see below, the descriptions of primitive accumulation in Section 6, like Sections 3 and 5, are descriptions of the various elements of primitive accumulation rather than the primitive accumulation itself.

First, the paragraph following the paragraph containing the quotation at the end of the previous section begins with the following sentences containing the word "primitive accumulation." "The discovery of gold and silver in America, the extirpation, enslavement and entombment in mines of the aboriginal population, the beginning of the conquest and looting of the East Indies, the turning of Africa into a warren for the commercial hunting of black-skins, signalized the rosy dawn of the era of capitalist production. These idyllic proceedings are the chief momenta of primitive accumulation. On their heels treads the commercial war of the European nations, with the globe for a theatre."

Afterwards, you will see sentences like the following:

"The different momenta of primitive accumulation distribute themselves now, more or less in chronological order, particularly over Spain, Portugal, Holland, France, and England. In England at the end of the 17th century, they arrive at a systematical combination, embracing the colonies, the national debt, the modern mode of taxation, and the protectionist system."

"The English East India Company, as is well known, obtained, besides the political rule in India, the exclusive monopoly of the tea-trade, as well as of the Chinese trade in general, and of the transport of goods to and from Europe. But the coasting trade of India and between the islands, as well as the internal trade of India, were the monopoly of the higher employés of the company. The monopolies of salt, opium, betel and other commodities, were inexhaustible mines of wealth. The employés themselves fixed the price and plundered at will the unhappy Hindus. (Omission) Great fortunes sprang up like mushrooms in a day; primitive accumulation went on without the advance of a shilling." (pp. 752-753)

"But even in the colonies properly so called, the Christian character of primitive accumulation did not belie itself. Those sober virtuosi of Protestantism, the Puritans of New England, in 1703, by decrees of their assembly set a premium of £40 on every Indian scalp and every captured red-skin: in 1720 a premium of £100 on every scalp:" (p. 753)

"The public debt becomes one of the most powerful levers of primitive accumulation. As with the stroke of an enchanter's wand, it endows barren money with the power of breeding and thus turns it into capital, without the necessity of its exposing itself to the troubles and risks inseparable from its employment in industry or even in usury." (pp. 754-755)

"With the national debt arose an international credit system, which often conceals one of the sources of primitive accumulation in this or that people." (p. 755)

"Liverpool waxed fat on the slave-trade. This was its method of primitive accumulation." (p. 759)

It can be seen that the "primitive accumulation" that appears in Section 6 above includes various elements that cannot be captured by the "process of separating producers and means of production." The colonial system, the national debt system, the modern tax system, the protectionist trade system, these elements form a mercantilist policy system aimed at the additional accumulation of monetary wealth. "Capital" refers to these as "the chief momenta of primitive accumulation." However, if we think about it in retrospect, all of these elements are related to the latter half of Marx's critical system of economics, such as the state, foreign trade, and the world market. "Capital" itself is thought to be primarily concerned with the first half of the system related to capital, land ownership, and wage labor.[35] Although primitive accumulation, which is a historical process, is a uniquely national process, it is necessary to discuss it while also taking into account the international relationships surrounding it. From this perspective, it can be inferred that in this section of "Capital," the depiction of the primitive accumulation period "crosses the border" into the latter half of the system for criticism of economics.[36] However, the relationship between these mercantilist policies and the inherent primitive accumulation process (the separation process between producers and the means of production) is not explored here at all. This is probably one of the tasks of the latter half of systematic theory that Marx himself had planned but never fulfilled.

6. Conclusion — "Primitive Accumulation" Per Se and "Elements of Primitive Accumulation"

In this chapter, in relation to the concept of primitive accumulation developed in the current Volume 1, Chapter 24 of "Capital," in response to Mr. X's criticism of my theory, I discussed Marx's concept of primitive accumulation, that is, "the so-called primitive accumulation, therefore, is nothing else than the historical process of divorcing the producer from the means of

35 Regarding the "plan question" regarding the extent to which "Capital" encompasses Marx's plan on a system that criticizes economics in the late 1850s, see Sato et al. (1977). As stated in the text, I consider "Capital" to be the embodiment of the first half of this plan (capital, land ownership, and wage labor).

36 "Crossing the border" to the latter half of the system can be seen throughout "Capital." To describe the main parts, in addition to the parts mentioned in this section, Volume 1 includes Chapter 3, Section 3, c, "Universal Money," Chapter 20, "National Differences of Wages," Chapter 25, "Modern Theory of Colonization." Also, Volume 3, Chapter 35, "Precious Metal and Rate of Exchange."

production." Mr. X's argument was as follows: "Primitive accumulation has the aspect of creating a productive working class and, on the other hand, the accumulation of monetary wealth that is converted into capital. Your argument ignores the latter aspect." The difference between the two lies in whether or not the element of "accumulation of monetary wealth" is included in the concept of primitive accumulation. The "inclusion" argument seemed to have its roots in the confusion in Rosenberg's "Commentary on Capital" (1931-1933). In this respect, Mr. X's view was an echo of Rosenberg's view, which echoed almost 90 years later. I have considered the commentaries that I believe are representative of each era in Japan since the 1960s. As a result, the argument for including "accumulation of monetary wealth" in the tasks in Chapter 24, which was the source of Rosenberg's confusion, receded from the beginning to the background. In the 1980s, when the "System of Capital" was published, the issue of the Chapter 24 was to elucidate the "process of separation between producers and the means of production." Furthermore, even in Rosenberg, there is no argument to include "accumulation of monetary wealth" in the concept of primitive accumulation itself. It belongs to Mr. X.

However, even when considered in this way, the question of how the initial "monetary wealth" that gave rise to capitalist society was accumulated remains. This chapter provided the answer to this question: "It already existed during the primitive accumulation period."

By the way, one of the reasons why it is so difficult to understand the aforementioned concept of primitive accumulation from "Capital" is that Chapter 24 does not just develop the concept of primitive accumulation. Chapter 24 has also a polemical significance in criticizing the economics of ownership, which emphasizes the "accumulation of monetary wealth." This has given rise to the misunderstanding that "Capital" itself emphasizes the "accumulation of monetary wealth." Furthermore, in Chapter 24, the social conditions at home and abroad that encompass the historical process of primitive accumulation are described as the various elements of primitive accumulation. This can give rise to the mistaken idea that these conditions are a description of primitive accumulation itself. What I think it is fair to claim as my own unique opinion here is that I have clearly distinguished and shown two words that Marx had in fact distinguished as different things. The two words are the primitive accumulation itself (that is, its definition) and the elements of the primitive accumulation.

I think that the distinction between the concept of primitive accumulation itself and the various elements of primitive accumulation is particularly important when applying this concept to countries other than the England. The description of the period of primitive accumulation in "Capital" is particularly British, and this is partly reflected in the description of the various elements of primitive accumulation. On the other hand, in countries other than the England, there are likely to be different elements of primitive accumulation, and the specific aspects of the primitive accumulation process are also different. Only by firmly holding the concept of "separation of producers and means of production" will it be possible to understand the differences in the nuances of the process itself between countries. On the other hand, it also makes it possible to recognize the national diversity that can be seen in the various elements of primitive accumulation. Furthermore, it also becomes possible to grasp the uniqueness of each country's primitive accumulation period, which is determined by these two factors.

Chapter 4 : Community as a Social System

1. Theme

The world economy exists as a coexistence of areas[37] where capitalist relations are dominant and areas where non-capitalist relations are dominant, and as a combination of the two.[38] It goes without saying that the capitalist areas and the non-capitalist areas do not coexist and combine in a static equilibrium. There is a historical dynamic in which the former is constantly penetrating the latter, and as a result, the latter is being transformed under the influence of the former. A clear example of this situation today is the process of economic development in developing countries. This is because capital relations are often introduced exogenously, rather than naturally, into a society in which non-capitalist relations predominate. We can also see a process in which the capital relations take root in the developing countries and develop.[39] This kind of relationship between production modes in which one brings about a transformation in the other while combining is called "articulation." [40]

By the way, production relations in non-capitalist areas are often characterized by the existence of communities, and the fact that these

37 In this case, an "area" is a "social structure" consisting of a layered structure of multiple production modes, with spatial and geographical elements taken into account. Since economic activities are carried out using land as its basic foundation, a certain mode of production that is dominant there necessarily corresponds to a particular land space inhabited by humans.

38 "The world market exists not as a market in general, but only as a world economy that combines the world market and non-market society." (Sasaki 1993)

39 This vision of developing countries' development is presented by the theory of new international division of labor. (Fröbel et al. 1980)

40 Regarding the concept of "articulation," see Mochizuki (1981a).

communities have a relatively strong position as the basis of society, with varying degrees of intensity depending on the country or region.[41] Therefore, the process of creation and development of capitalist relations is often predicated on the dissolution of the community,[42] or its content is the gradual dissolution of the community.[43] Capitalist relations can easily subsume non-capitalist relations because their basic form, the commodity form, has an external and superficial character relative to social production. In some cases, the commodity economy gradually penetrates into a community based on a self-sufficient economy, which becomes an opportunity to differentiate and disintegrate its homogeneous members, and undermines the community from within. In other cases, existing non-capitalist class relations within a community are reorganized to accommodate the penetration of the commodity economy, which destroys the communal order. The role of the state will be to bring a money economy into the subsistence economy through taxation and thereby facilitate these processes. However, although these often become the dominant aspect of the situation that actually occurs with the penetration of the commodity economy, it is still only

41 Otsuka (1955) characterizes pre-capitalist societies as ones in which communities exist extensively. In his book, the basic unit of human groups that control land in pre-capitalist societies is called a community, and my terminology also follows this. Regarding the broader scope of Max Weber's concept of community, see Uchida (1996). Matsuo (1978) examines the process by which a pre-capitalist class society was established based on Otsuka's theory of community.

42 "The relationship of labor to capital or to the objective conditions of labor as capital, presupposes a historical process which dissolves the different forms, in which the laborer is an owner and the owner labors. This means first and foremost: (1) a *dissolution* of the relation to the earth - to land or soil - as a natural condition of production which man treats as his own inorganic being, the laboratory of his forces and the domain of his will. All forms in which this property is found, assume a communal entity whose members, whatever the formal distinctions between them, are *proprietors* by virtue of being its members. Hence, the original form of this property is *direct communal property*." (Marx 1858)

43 It was Luxemburg (1913) who portrayed capitalist accumulation as inseparable from the process of dismantling the non-capitalist areas.

one aspect. This is because communities do not disintegrate without resistance to the pressures of the market and the state, but rather maintain the autonomy to maintain their unique characteristics while more or less defending themselves against the pressures.[44] Alternatively, communities try to survive while adapting to the commodity economy.[45] In this way, in many developing countries, communities exist relatively strongly compared to areas of advanced capitalism, although there are variations depending on the country or region.[46] Furthermore, communities are in conflict with the commodity economy while exerting influence within society.[47] Therefore, violence by the ruling class is often used to destroy communities.

However, what this chapter is concerned with is not this process of conflict. Rather, the question is going back further and asking what kind of logic is inherent in a community that defines its unique character. There are already some classics on this subject that we should study, and we will revisit as

---

44 From what has been said here, the question arises regarding the types and categories of commodity economies, such as what kind of commodity economy leads to the dissolution of a community, and what kind of commodity economy conversely strengthens a community. Otsuka (1962) discussed this issue by focusing on the difference between the intra-community division of labor and the inter-community division of labor.

45 Engels (1884) suggested the adaptation of the community to the capitalist economy. Meillassoux (1975: Part 2) also focused on West Africa and explained the specific mechanism of this adaptation from a labor market perspective.

46 Regarding this variation, in addition to the morphological differences in the community shown in "Pre-Capitalist Economic Formations" and "Reply to V. I, Zasulich's Letter," there is also the transformation of the community into a secondary social structure, and the transformation of the community due to the penetration of commodity economy into the community.

47 Viewing the conflict between the community and the commodity economy at the micro level has sparked a debate over whether the behavior of farmers in developing countries is "rational" or "emotional." Takahashi (1999) provides a concise summary of this debate that has been taking place in Southeast Asia. For information on Africa, see Sugimura (2004: Chapter 2). Scott (1976), for example, describes this "conflict" in colonial Southeast Asia that eventually led to peasant revolt.

needed them in this chapter. In particular, Marx's series of works on community, with the two peaks of "Pre-Capitalist Economic Formations" ("Formations") and "Reply to V. I, Zasulich's Letter" ("Reply"),[48] it can be said that they have continued to have an extremely strong influence on the understanding of community in Japan, including those who criticize them. For example, Hisao Otsuka's "Basic Theory of Community" (1955),[49] which is also a classic that continues to be read to this day, discusses various forms of community ownership, which is a leitmotif of "Formations," while also referring to "Reply." Therefore, this chapter will also proceed with an examination of Marx's community theory. However, for the reasons explained in the next paragraph, I will focus specifically on examining "Formations" among Marx's series of works. "Formations" was written as a manuscript around 1858, about 10 years before the publication of Volume 1 of "Capital," and the first two-thirds of it was devoted to community, and the remaining third is devoted to a description of primitive accumulation.

However, what is the need to reexamine the "Formations" that previous studies have been reading for many years? The reason for this is actually related to what I just mentioned, namely that Otsuka's community theory interprets "Formations" as a theory of forms of ownership, and reorganizes the logic of that work from that perspective. It is true that Marx's awareness of the problem in "Formations" posited capitalist society as one extreme, as a society in

48 There is a chronological gap of more than 20 years between the 1881 "Reply" and its predecessor, "Formations." Therefore, the deepening of Marx's own study of communities during this period has resulted in a discrepancy in the descriptions of the geographical and staged types of communities between the two works. Otani et al. (2013: Chapters 7 and 8) traced this process back to Marx's excerpted notes. A detailed text critique of Marx's community theory is by Kotani (1979, 1982). Meillassoux (1960: p. 39) also states that Marx and Engels focused on the economic laws unique to primitive societies. Taguchi (1979: Chapter IV) provides an interesting review of the Meillassoux's paper.

49 A recent reexamination of Otsuka's book is by Onozuka et al. (2007).

which private ownership was widespread, while positioning the various forms of community as a logical progression that led to that point.[50] While respecting Marx's awareness of this problem, let us elaborate on the description in the draft "Formations." As a result, it is quite natural to arrive at Otsuka's theory of community, which focuses on the "inherent duality" of communal occupation and private occupation, which is inherent in a community, and focuses on forms of ownership. However, if we step back a little from the leitmotif of "Formations" and think about how community theory should be in the first place, it becomes clear that community theory cannot be reduced to a theory of ownership forms. This corresponds to the fact that theory of capital cannot be reduced to theory of capitalist ownership. Just as the theory of capital requires a theory of internal laws of motion that is developed on the basis of capitalist property relations, so too is the theory of community necessary to elucidate its internal logic that is developed on the basis of community ownership. However, it is not the task of this small chapter to fully develop the internal logic of the community. Here, I will limit my task to exploring the point of view. In other words, what I am concerned with in this chapter is the internal logic of "community in general," which is abstracted from the differences in types and stages of community.[51]

---

50 After critically examining Otsuka's community theory, Fukutomi (1970) states the following: "Otsuka revised the concept of multisystem development in 'Formations' to 'a diagram of the development of three communities in one lineage;' Asian, Greco-Roman, and Germanic." However, it is questionable whether Marx had a clear concept of "multisystem development" at that time. Although the three forms were not thought to be in a relationship of unilinear, step-by-step transition in the same region, as I wrote in the main text, they were thought to be at least in a relationship of logical development. Fukumoto (2003) organized the controversy surrounding the historical laws of the Marxist school, focusing on how the concept of Asian modes of production has been understood.

51 The internal logic of the "community in general" must be a rule that applies to the community, which is, on the one hand, a category of economic history, and, on the other hand, a category for analyzing the current situation in developing countries. The idea of using the economic history category to analyze the current situation in developing countries was proposed by Akabane (1971).

Despite its basic character as a theory of ownership forms as mentioned above, "Formations" seems to contain useful descriptions for gaining points of focus when unraveling the internal logic of "community in general." [52] In Section 6 of this chapter, I will summarize the results of a rural survey conducted by the author in the Inner Niger Delta beneath the Sahara Desert, in order to illustrate the analytical perspective on the internal logic of community as described below.

2. Structure and Method of the Community Theory in "Pre-Capitalist Economic Formations"

An examination of the theory of community developed in the earlier sections of "Formations" begins here with a consideration of its structure and method. The structure is that the theory characterizes each form in the order of Asiatic, Roman, and Germanic, and subsequently discusses slavery and serfdom. However, in the way the three formations are discussed, there are quite large differences between the Asian form being explained at the beginning and the other Roman and Germanic forms. The first Asian form is discussed in its own right, as it were. On the other hand, the Roman and Germanic forms are depicted as opposites, where private property as a negation of community is being generated within the community. However, the complete opposite of the community is a capitalist society in which private property has come to cover all aspects of society, and the Roman and Germanic forms are positioned as transitional formations leading to that point.

These characteristics of argumentation methods also influence the content of arguments regarding each formation in the following ways. That is, in the first description of the Asian form, the characteristics of the Asian form as well

52 In this sense, this chapter is the work of "reevaluating Otsuka's theory of community based on a Marxian perspective." (Onozuka et al. 2007)

as the definition of "community in general," which is the subject of this chapter, are included, although the distinction between the two is not sufficiently clear. On the other hand, regarding the characteristics of later Roman and Germanic forms, descriptions of differences and contrasts with other formations, including Asian form, are conspicuous. What makes these differences in writing possible is the typicality of the Asian form, which best embodies the characteristics of the "community in general." The following text by Marx discusses the typicality of Asian form (Oriental form).

"the oriental form, modified among the Slavs; developed to the point of contradictions in classical antiquity and Germanic property, though still the hidden, if antagonistic, foundation." (Marx 1858)

Therefore, in the abstraction of the characteristics of Asian forms, which are emphasized in contrast to Roman and Germanic forms, there are contents related to the internal logic of the community in general.

Let us begin this work by understanding the definition of Asian form. Previous researchers have focused on the powerful attraction of the weighty phrases in "Formations" descriptions of Asian form. They have often focused their attention on understanding what have been considered to be its characteristic features (tyranny and general slavery) compared with other forms. On the other hand, it seems that the internal logic of the community in general, which is included in these descriptions, has not always been sufficiently read into.

3. Definition of the Asian Form

The description of the Asian form of community at the beginning of "Formations" is long, but it is important for this chapter, hence I will quote it below.

"For instance, as is the case in most Asiatic fundamental forms, it is quite compatible with the fact that the *all-embracing unity* which stands above all these small common bodies may appear as the higher or *sole proprietor*, the real communities only as *hereditary* possessors. Since the *unity* is the real owner, and the real precondition of common ownership, it is perfectly possible for it to appear as something separate and superior to the numerous real, particular communities. The individual is then in fact propertyless, or property – *i.e.*, the relationship of the individual to the natural conditions of labor and reproduction, the inorganic nature which he finds and makes his own, the objective body of his subjectivity – appears to be mediated by means of a grant [*Ablassen*] from the total unity to the individual through the intermediary of the particular community. The despot here appears as the father of all the numerous lesser communities, thus realizing the common unity of all. It therefore follows that the surplus product (which, incidentally, is legally determined in terms of [*infolge*] the real appropriation through labor) belongs to this highest unity. Oriental despotism therefore appears to lead to a legal absence of property, in most cases created through a combination of manufacture and agriculture within the small community which thus becomes entirely self-sustaining and contains within itself all conditions of production and surplus production.

Part of its surplus-labor belongs to the higher community, which ultimately appears as a *person*. This surplus-labor is rendered both as tribute and as common labor for the glory of the unity, in part that of the despot, in part that of the imagined tribal entity of the god. In so far as this type of common property is actually realized in labor, it can appear in two ways. The small communities may vegetate independently side by side, and within each the individual labors independently with his family on the land allotted to him.

(There will also be a certain amount of labor for the common store – for insurance as it were – on the one hand; and on the other for defraying the costs of the community as such, *i.e.*, for war, religious worship, etc. The dominion of lords, in its most primitive sense, arises only at this point, e.g., in the Slavonic and Rumanian communities. Here lies the transition to serfdom, etc.)

Secondly, the unity can involve a common organization of labor itself, which in turn can constitute a veritable system, as in Mexico, and especially Peru, among the ancient Celts, and some tribes of India. Furthermore, the communality within the tribal body may tend to appear either as a representation of its unity through the head of the tribal kinship group, or as a relationship between the heads of families. Hence, either a more despotic or a more democratic form of the community. The communal conditions for real appropriation through labor, such as irrigation systems (very important among the Asian peoples), means of communication, etc., will then appear as the work of the higher unity – the despotic government which is poised above the lesser communities." (Marx 1858)

The contents regarding the Asian form of community included in this quotation are organized in my own way and broken down into bullet points as follows.

1) It is consistent with the basic Asian form that the "all-embracing unity" embodied in a despot appears as the "higher or sole proprietor" over some communities.

2) In contrast to the "all-embracing unity," the communities appear only as "hereditary possessors."

3) Ownership of the "all-embracing unity" is transferred to its members through the community.

4) Under such ownership relationships, surplus-labor belongs to the "all-embracing unity."

5) In the case of despotism, community property actually exists in the midst of "a legal absence of property." What creates this kind of community

ownership is "a combination of manufacture and agriculture within the small community." As a result of this combination, "the small community becomes entirely self-sustaining and contains within itself all conditions of production and surplus production."

6) The forms in which surplus-labor belongs to the "all-embracing unity" include "tribute" of surplus products and "common labor for the glory of the unity."

7) In reality, there is community property that is "actually realized in labor," but it has the following structure. ① The "small communities" live independently and coexist with each other (as mentioned above in 5). ② If we look within the community, an individual may work independently with his or her family on the allotted land. ③On the other hand, there is labor for "the common store" and "insurance," and labor to cover communal expenses, such as for war, religious worship, etc. From this, "the dominion of lords, in its most primitive sense 'appears' e.g., in the Slavonic and Rumanian communities." Moreover, "here lies the transition to serfdom, etc."

8) Labor may be fully communalized within a community. Sometimes this becomes a "veritable system," as in the case of Mexico, Peru, the ancient Celts, and some Indian tribes.

9) The "all-embracing unity" may be represented by "the head of the tribal kinship group, or as a relationship between the heads of families." And depending on which one it is, the form of the community is determined to be "more despotic or more democratic."

10) "The communal conditions for real appropriation through labor, such as irrigation systems (very important among the Asian peoples), means of communication, etc., will then appear as the work of the higher unity – the despotic government which is poised above the lesser communities."

I think it is possible to further divide Marx's descriptions organized in this way into the following two categories; ownership items and surplus-value acquisition items. Note that 9) is a definition of the character of the "all-embracing unity," and it is not forced to classify it into one of these two item groups.

(A) Items of Ownership

These apply to 1) 2) 3) 5) 7) 8) 10) above. If I expand on those descriptions in my own way, it becomes as follows.

In the Asian form of community, the natural conditions of production, consisting of land and its appurtenances, may appear as the property of the "all-embracing unity" who is only one legal owner (i.e. if there is the "all-embracing unity"). However, in this case as well, ownership in its fundamental sense, that is, ownership as a relational act to the natural conditions of production,[53] is multilayered, as will be seen below. In the following parts of this section, ownership at lower levels will be referred to as "possession," to distinguish it from the highest level of ownership by the "all-embracing unity."

First, there are two cases regarding the labor of the individuals who make up the community: (a) By joint labor on common land. (Marx 1867) (b) Working independently with their families on allotted land. In case (b), the individuals become possessors of the allotted land. Occupation of allotted land by individuals may be hereditary or non-hereditary.[54] If (a) is primary, there is no question that individuals are dependent on the community. On the other hand, even when (b) is the main factor, the lives of individuals do not end at

53 "Thus originally *property* means no more than man's attitude to his natural conditions of production as belonging to him, as the *prerequisites of his own existence*;" (Marx 1858)

54 "Where property exists *only* as communal property, the individual member as such is only the *possessor* of a particular part of it, hereditary or not," (Marx 1858)

the family level for the following reasons, and as a result, individuals are forced to live within a communal spatial extension.

The first reason is that agriculture and industry are combined within communities, resulting in the existence of self-sufficient and self-contained living areas at the community level. If we look at this from the perspective of individuals, we can see that their lives are completed only within the spatial expanse of a community. In this way, a self-sufficient living area exists within the community based on the division of labor between agriculture and industry. Under these circumstances, a situation emerges in which the individuals who make up the community as a whole interact with the natural production conditions represented by the land, and engage in material metabolism with them. Because of this strong dependence of individuals on the community and the self-sufficient character of the community, "Formations" later asserts the "robustness" of Asian form as follows.

"The Asiatic form necessarily survives the longest and most stubbornly. This is due to the fundamental principle on which it is based – that is, that the individual does not become independent of the community; that the circle of production is self-sustaining, unity of agriculture and craft manufacture, etc." (Marx 1858)

By the way, this issue of the robustness of the Asian form based on the integration of agriculture and industry within a communal space was later discussed in "Capital," Volume 3, Chapter 20, "Historical Facts about Merchant's Capital," as following paragraph. When explaining the resistance shown by the Indian community to the innovative forces of British commerce, the specific impetus of "savings on '*faux frais*' of the circulation process" is referenced.[55]

55 The Indian cotton spinning industry sprang up in the 1860s, triggered by the "cotton famine" caused by the American Civil War (1861-1865) and the boom in

"In India the English lost no time in exercising their direct political and economic power, as rulers and landlords, to disrupt these small economic communities. English commerce exerted a revolutionary influence on these communities and tore them apart only in so far as the low prices of its goods served to destroy the spinning and weaving industries, which were an ancient integrating element of this unity of industrial and agricultural production. And even so this work of dissolution proceeds very gradually. And still more slowly in China, where it is not reinforced by direct political power. The substantial economy and saving in time afforded by the association of agriculture with manufacture put up a stubborn resistance to the products of the big industries, whose prices include the *faux frais* of the circulation process which pervades them." (Marx 1894)

However, the fact that agriculture and industry come together to form a self-sufficient sphere within a relatively narrow space is not unique to Asian form. Rather, it is likely to be commonly found in underdeveloped economies.[56] However, because individual communities exist isolated and dispersed in a vast space, the human and spatial scope of such integrated agricultural and industrial self-sufficiency sphere overlaps with the scope of the community, and as a result, there is a strong dependence on the community of members. This is probably what Marx is discussing as a characteristic of Asian form.[57]

Indian cotton as an alternative supply. Local traders who made huge profits from cotton trading during the boom invested in the machine spinning industry. Until the 1870s, its development was driven by domestic demand, supplying thick yarn to domestic handloom weavers, but around the 1880s, exports to China and Japan began. In the 1890s, it went so far as to drive out British cotton yarn in the Chinese market. (Akita 2012)

56 "Domestic handicrafts and manufacturing labor, as secondary occupations of agriculture, which forms the basis, are the prerequisite of that mode of production upon which natural economy rests — in European antiquity and the Middle Ages as well as in the present-day Indian community, in which the traditional organization has not yet been destroyed." (Marx 1894)

57 Engels (1875) describes the complete isolation of individual communities from

Second, people are not free from the vulnerability of frequent natural disasters when their productivity is at a low level of development, but saving in preparation for such natural disasters is necessary for survival. For the following reasons, it is difficult to stockpile at the level of individual member families under low productivity, therefore a joint effort is necessary.

A portion of the surplus of products that exceeds the immediate means of subsistence will be appropriated for stockpiling, but if the development of productivity is low, this may swallow up almost all of the surplus. At the same time, the conditions for forming such a surplus are not uniformly given to all families. This is because production conditions such as the land conditions (fertility and area) of the allotted land and the family labor force usually differ to a greater or lesser extent among families. Alternatively, even if we assume that the land conditions of allotted land are exactly the same, how the land is used for production may differ between families.

The agricultural system of medieval Germany, characterized by the *Hufe* system and communal regulation (Weber 1924), and the similar agricultural system of medieval France with *mansus* and communal regulation (Bloch 1931). It is true that these systems, as long as they remain close to the ideal model, appear to be aimed at equalizing differences between families in terms of ownership and management as much as possible and achieving equality in production results. Furthermore, this orientation towards equality of production results is likely to be widely seen in other forms of communities, although the specific mechanisms may change.

However, to think that such an equalization mechanism is fully functioning and always achieves its purpose is to place excessive expectations on the

---

each other as the natural basis of Oriental despotism. A similar description of the relationship between community isolation and tyranny is given by Marx (1881). Vidal de la Blache (1922), a French geographer, also discusses the "isolation" and "segregation" of communities in India.

system, and is, in fact, an illusion. And in the same year, it will actually happen that some families will have a surplus, while others will not. Some families may even find themselves short of immediate necessities in the same year.

Thus, stockpiling requires a collective effort because there can be significant differences between families in the conditions that produce surpluses and the results of production activities. In other words, a family that happens to be blessed with the conditions to produce a surplus in a given year can contribute to the community's stockpiling, but a family that is not blessed with such conditions cannot contribute to stockpiling. On the contrary, some of the latter may manage to maintain their livelihood by drawing down community reserves. Therefore, stockpiling has the function of redistributing wealth from those who contributed to stockpiling to those who could not. Moreover, rather than appropriating surplus-labor and surplus products for the enrichment of individuals, they are appropriated for the stockpiling of the community. In order for this to be realized smoothly, administrative mechanisms and religious practices must exist to absorb these surpluses into the community. And there must be a system of laws and ideologies that legitimize these things while also viewing the pursuit of private interests, which is the exact opposite, as an evil that should be suppressed.[58] Therefore, no matter how irrational these systems of law and ideology may seem at first glance, they have a rationality that allows society to survive under low productivity. In this sense, Lévy-Bruhl's statement (1910), which attempts to explain the systems and activities of pre-capitalist communities based on the "irrationality" of its members, is probably premature. "It is a waste of effort to try to explain the institutions, customs, and beliefs of primitive people based on

58 "Religion affords such strong motives to the practice of virtue, and guards us by such powerful restraints from temptation of vice, that many have been led to suppose, that religious principles were the sole laudable motives of action." (Smith 1759)

psychological and logical analysis of the 'human spirit' as seen in our society. Interpretation of them is insufficient unless we take as our starting point the pre-logical, mystical mentality on which the various forms of activity of these primitive peoples depend."

An example of such an ideology, albeit somewhat abruptly, can be found in Aristotle (edited around 300 BC). In his theory, the use of "wealth" consists of consumption and gifting, and a "generous" person who possesses the virtue of moderation gives the right things to the right people, at the right time, according to his financial resources. Moreover, he consumes comfortably. And he takes from what is due, and in what is due. A "generous" person respects "wealth" not for its own sake, but for the sake of what it gives. The vices that oppose "generousness" are "waste" and "stinginess." "Generousness," along with other virtues such as "bravery," "temperance," and "moderation," constitutes "justice" in a broad sense, and is the basis of the law that orders such actions.

By the way, one method of allocating surplus to the community's reserves is when the labor that produces the surplus itself is done as communal labor by those who are able to contribute (common labor). As mentioned earlier, the status of these people who can serve may differ from family to family. The second is a case where each family performs the labor separately on the shared land, but there is a mechanism in place for the surplus product generated from that work to be absorbed into the community (through the donation of surplus). Incidentally, in the latter case, assuming an uncivilized society where the degree of development of productive forces is extremely low, neither the conditions of production under which surplus is formed nor the conditions under which surplus is not formed are fixedly tied to any particular family. There is a possibility that almost everyone will go hungry,[59]

59 Sahlins (1972: Chapter 2) emphasizes this point.

so there is always the possibility that the relationship between rich and poor will be reversed. In other words, the lower the degree of development of productive forces, the more fluid the hierarchical relationships within the community will be. Therefore, in these primitive societies, the majority of people may benefit from the surplus being appropriated for stockpiling, and the stockpiling function therefore appears to provide happiness to everyone. On the other hand, since the surplus is absorbed into the community, it cannot be stored at the family level, and the relationship between the rich and the poor is not fixed but becomes fluid, which is the opposite causal relationship. In other words, there is a mutually determining relationship between the stockpiling function of the community and the mobility of rich and poor. However, in primitive societies, due to the low level of productivity, even if a family were to save up surplus, there are limited conditions under which it can be used to fund the development of infrastructure and the introduction of new technology to develop productivity. Therefore, there is also little room for the relationship between the rich and the poor to be fixed through the accumulation of surplus.[60] Therefore, it can be said that the fluidity of rich and poor and class relations within a community is ultimately determined by the state of extremely low productive power in primitive societies.

Since there is only a very simple description of the stockpiling function of the community in "Formations," my elaboration from there may be criticized as going too far. In response to such criticism, it would be helpful to provide a concrete description of the community's stockpiling function. The first thing I would like to introduce is a quote from Cunow's "Complete History of Economy" (1926-1931). The book contains scattered descriptions of communal

60 "it seems to me that men cannot live conveniently where all things are common. How can there be any plenty where every man will excuse himself from labor? for as the hope of gain doth not excite him, so the confidence that he has in other men's industry may make him slothful." (More 1516)

storage facilities observed in North American Indian villages, but here I will cite the most detailed description of one tribe. Since the subject of this section is Asian form, some readers may feel uncomfortable about using North American Indian communities here as illustrative material without making any argument as to whether or not they are Asian form. However, as I will discuss later in Section 5, I envision the stockpiling function as something that is inherent in communities in general, rather than as something unique to Asian form. I think so, even if it may take the form of a secondary and distorted form, such as the protection of the people by the feudal lord. If this is the correct view, it would be permissible to seek examples to visualize the concrete form of the stockpiling function without limiting it to Asian form.

"In addition to the outer field distributed among members of the same clan, village had common outer field. It was not subdivided according to the number of kin, but was the property of the entire village community and cultivated by it jointly. The proceeds were collectively channeled into a common depot, from which aid was given to fellow villagers in times of crop failure, floods, and war. Charles C. Jones called this house the 'general store house,' and said of it: 'It was built and cared for by the collective labor of the tribe. In it, corn and various fruits were stored, as well as the meat of fish, deer, crocodiles, snakes, dogs, and other animals, which had been smoked and hung on scaffolding to dry.' Joan Bertram, on the other hand, calls this village depot 'Kings cribs' because it is under the supervision of the village chief. He describes it as follows: 'Such communal repository serves the wisest and best purposes. In other words, it was a storage warehouse or rescue method in case of emergency. When a family's treasury became poor, in case of misfortune or emergency they (the Indians) turned to the king and received support and wages from the treasury. It also served to assist other settlements when they were suffering from lack of supplies, and to feed troops, travelers, and visitors from elsewhere.'"

As we can see, what Cunow describes here is "through common labor" according to the previous classification of the mechanism for absorbing surplus into the community.

An example of the other "through the donation of surplus" can be found in the "provisions regarding one-tenth of the harvest" in the Old Testament. In other words, Deuteronomy Chapter 14, 28 and 29, says: "At the end of every three years you shall bring forth all the tithe of your increase in the same year, and shall lay it up within your gates: and the Levite, because he has no portion nor inheritance with you, and the foreigner living among you, and the fatherless, and the widow, who are within your gates, shall come, and shall eat and be satisfied; that Yahweh your God may bless you in all the work of your hand which you do."

The word "foreigner" appears here, and according to Weber (1920), "foreigners" in the Old Testament were small livestock herders, craft workers, merchants, priests, wage laborers and musicians in ancient Israel. They were guest clans who were outside the community and did not own land but were legally protected. The "provisions concerning one-tenth of the harvest," the "gleaning law," and the "sabbatical year," which we will discuss shortly, also protect these people. This means that the scope of application of these social protection provisions went beyond the boundaries of individual communities. Its scope of application was the "community of Israel" as sectarian comrades, including those who lived in the gaps between communities, but it was also limited to that.[61]

Furthermore, "through the donation of surplus" is considered to correspond to the redistribution of the three forms of integration when the economy is

61 Regarding the dualism of internal and external morality in the "community of Israel," see Weber (1920).

institutionalized, namely reciprocity, redistribution, and exchange, as discussed by Polanyi (1957). Polanyi discusses redistribution as follows.

"Redistribution within a group occurs as long as the distribution of goods is concentrated in one hand and is done by custom, law, or central decisions. Sometimes it is a physical levy followed by storage-redistribution, and sometimes the 'levy' is not physical but merely appropriative. In other words, it may simply belong to the right to dispose of the physical location of the good. Redistribution occurs for many reasons and occurs at all levels of civilization, from primitive hunting tribes to the great storage systems of ancient Egypt, Sumer, Babylonia, and Peru."

Polanyi envisions a community economic system as a subsistence economy supplemented by a market, in contrast to a capitalist economic system that is integrated almost exclusively by exchange through the market. In addition to the redistribution we have just seen, he is also considering another form of material circulation that does not depend on exchange: reciprocity, such as gifts and mutual aids between symmetrical groups. In a primitive society where the relationship between rich and poor is fluid, the relationship between individuals is symmetrical, and survival is therefore guaranteed through reciprocity.[62] Alternatively, as stratification progresses within a community, reciprocity transforms into a patron-client relationship, where lower classes respond to the protection provided by upper classes with service and loyalty.

Examples of the redistribution of goods, which appears in Polanyi's current discourse and which simply belongs to the right to dispose of the physical location of goods, can be found in the Old Testament. Namely, the aforementioned "gleaning law" and the "sabbatical year." Regarding the "gleaning

62 Tamanoi (1978) characterizes communal exchange as follows, emphasizing the difference from capitalist exchange. "It must be understood that in a community, exchange was included in and defined by this collective act of reciprocity."

law," God commands the following in Leviticus 19, 9 and 10: "When you reap the harvest of your land, you shall not wholly reap the corners of your field, neither shall you gather the gleanings of your harvest. You shall not glean your vineyard, neither shall you gather the fallen grapes of your vineyard; you shall leave them for the poor and for the foreigner." Regarding the "sabbatical year," it is written in Exodus 23, 10 and 11 as follows: "For six years you shall sow your land, and shall gather in its increase, but the seventh year you shall let it rest and lie fallow, that the poor of your people may eat; and what they leave the animal of the field shall eat. In like manner you shall deal with your vineyard and with your olive grove." [63] Deuteronomy 23, 25 and 26 contains the following detailed regulations regarding how to eat when a hungry person enters someone else's field or vineyard to eat. "When you come into your neighbor's vineyard, then you may eat of grapes your fill at your own pleasure; but you shall not put any in your vessel. When you come into your neighbor's standing grain, then you may pluck the ears with your hand; but you shall not move a sickle to your neighbor's standing grain."

Finally, returning to the example of North American Indians, let us now look at Morgan's (1881) statement.

〈Among the Iroquois tribes, members of household cultivated the fields, gathered the harvest, and stored it in their dwellings as a communal reserve. However, to a greater or lesser extent, products were owned by individuals or by individual families. For example, peeled corn was tied into bundles with its husks and hung in each room. But when one family ran out of provisions, other families provided as much food as they could store. Friends engaged in hunting and fishing also shared their catch. After returning home, the surplus

63 According to Weber (1920), the provision of the "sabbatical year" was a "moral provision" derived from religious sermons, and was not a "legal command" based on statute law. However, it had a practical influence in the development of Judaism.

was distributed among the families in each household, preserved, and stored for winter food. Villagers do not store food together and do not share food unnecessarily. Distribution was limited to households. However, when people were in dire straits, a helping hand was finally extended out of the custom of hospitality.〉

In other words, for the Iroquois, the unit of communal labor and storage was not a village, as in the case of Cunow, but a "household," a group of blood-related families. On the other hand, villages did not share food storage. However, when a "household" fell into poverty, it is said that the "households" helped each other through the "custom of hospitality." What is the "custom of hospitality?"

〈It has been a custom among the Iroquois tribes since time immemorial to extend hospitality to any guest. In any Indian village, when someone came into the house, whether a villager, a tribesman, or an outsider, it was the woman's job to offer him food. Neglecting this role is tantamount to public humiliation. Guests who visited were expected to eat if they were hungry, and even if they were not hungry, they were expected to take a bite of the food as a courtesy and thank the woman who prepared it. This was always done no matter what home he entered or what time of day he entered. This was supported by people's feeling that it was natural to do this as a habit.〉

〈The customs of hospitality practiced by American Indians eventually led to food equality. It was impossible for an Indian village or encampment to be prosperous as a whole, while for one corner of the same village or encampment to be starving or in poverty.〉

This "custom of hospitality" also applied to guests from outside the village, and even Europeans are said to have received this custom of hospitality wherever they arrived in North America. However, there is no doubt that the main guests were neighboring residents of the same village. It is also clear that those who are in a state of poverty cannot afford to entertain others, but can only serve as guests. As a result, the "custom of hospitality" led to the redistribution of reserves within the village.

Now, the third reason why individuals are forced to live within a community is that there needs to be communal labor, that is, labor for civil engineering projects such as the construction of "irrigation systems and means of communication, etc." However, since there are elements of this third moment that cannot be completed at the level of a small community, it would be appropriate to take a closer look at it in detail in the following description of "all-embracing unity." In any case, the above three moments lead to the creation of communal possession.[64] Since these three moments are essential for the survival of individuals, the existence of individuals presupposes the existence of a community and appears as something mediated by the community. Or, the appropriation of the natural conditions of production by individuals appears to presuppose their appropriation by the community.

Upon such appropriation of the natural conditions of production by individuals and communities, the "all-embracing unity" may appear as the "sole proprietor." In that case, ownership by the "all-embracing unity" is not merely legal,[65] but has substance as a relational act to the natural conditions of production. This is because, among the "Asian peoples," civil engineering projects such as the construction of "irrigation systems and means of communication, etc." sometimes exceed the scope of collective efforts by individual "lesser communities." In that case, these projects would be handled

64 "His relation to the *land* as *his own* [*als dem seinigen*] in virtue of the community, communal landed property, at the same time *individual possession* for the individual, or in such a manner that the soil and its cultivation remain in common and only its products are divided." (Marx 1858)

65 I learned the meaning of "legal" from Shinozuka (1974).

by the autocratic "all-embracing unity" overlying small communities.[66] There, the community's appropriation of the natural conditions of production presupposes and is mediated by the ownership of those conditions by the "all-embracing unity."

Although ownership by the "all-embracing unity" is seen here in the "most" regions of Asian form, it is nevertheless not spoken of as a universal phenomenon in these regions. It should also be noted that there are regional and ethnic limitations attached to the "all-embracing unity," and that it is considered to be an attribute unique to the ethnic groups that require large-scale civil engineering projects. Furthermore, in a later section where Marx (1858) talks about "the specifically oriental form," he says that "what exists is only *communal* property and private possession." In other words, the existence of ownership by the "all-embracing unity" is temporarily left out of view.[67] He subsequently goes on to say that "historic and local, etc., circumstances may modify the character of this possession in its relation to the communal property in very different ways, depending on whether labor is performed in isolation by the private possessor or is in turn determined by the community, or by the unity standing above the particular community," and focuses on the existence of the "all-embracing unity" as one of the various historical and regional variations that the mode of private occupation exhibits. Moreover, as we saw earlier, in 9), the character of the "all-embracing unity" itself is described as being more autocratic or more democratic, that is, as rich in

66 Wittfogel (1957) gives this system the names "hydraulic society" and "hydraulic civilization" in order to express and emphasize the "agricultural bureaucracy" character.

67 The following is a similar description. "In the Asiatic form (or at least predominantly so), there is no property, but only individual possession; the community is properly speaking the real proprietor – hence property only as *communal property* in land." (Marx 1858)

nuance.[68]

Well then, what exactly are the Asian ethnic groups that require major civil engineering projects? Regarding this, Marx (1853), accepting Engels (1853), pointed out the peoples of the "desert region" as follows.

"There have been in Asia, generally, from immemorial times, but three departments of Government; that of Finance, or the plunder of the interior; that of War, or the plunder of the exterior; and, finally, the department of Public Works. Climate and territorial conditions, especially the vast tracts of desert, extending from the Sahara, through Arabia, Persia, India, and Tartary, to the most elevated Asiatic highlands, constituted artificial irrigation by canals and water-works the basis of Oriental agriculture."

Although these texts were written several years before Marx wrote "Formations," his concept seen here does not contradict the content described in "Formations." In view of this fact, it would be fair to say that this concept was still maintained even at the time of writing "Formations." [69]

By the way, the way of ownership in the Asian form seen in (A) is connected to the way of acquiring surplus-value there, as we will see next.

---

68 Fukutomi (1969) argued that a typology of "Asian forms" is necessary.

69 This concept may be related to the scope of Marx's knowledge at the time. Regarding Marx and Engels' knowledge of primitive society at the time of writing "Formations," see Hobsbawm (1964). Furthermore, Shimizu (1942), who describes the development process of rural society in medieval Japan from the Heian period to the Sengoku period, states that irrigation was the trigger for village federations. Ishimoda (1946) also argues that irrigation in medieval Japan was a problem that was difficult to solve "within villages," as shown in the following passage, but it was a small regional problem rather than a national level problem. In other words, Ishimoda shares the same direction of understanding as Shimizu. "The geographical conditions of our country's rivers and plains do not necessitate a huge centralized state for irrigation water problems. The topography, in which small plains centered on capillary rivers form independent geographical units, suggests that irrigation problems can originally be solved within the scope of these small geographic regions."

(B) Items of Surplus-Value Acquisition

The surplus-value acquisition items fall under 4) 6) 7) above. The following is an expanded version of the content based on those descriptions. Note that 7) overlaps with "(A) Ownership Items," but the reason for including it in (B) is also explained below.

Regarding the relationship of acquisition of surplus-value, "Formations" explains based on the one hand the relationship of ownership. In other words, surplus-labor in the form of "tribute" of surplus products or "communal labor for the glory of the unity" belongs to the "all-embracing unity" which is the "sole proprietor."

However, on the other hand, 7) argues that the relationship of acquiring surplus-value is based on stockpiling and communal labor within the community. In such case, it goes further than property relation and refers to activities within the community. In other words, the "the transition to serfdom, etc." lies in the "dominion of lords" over the surplus-labor performed within the community ("a certain amount of labor for the common store - for insurance as it were - on the one hand; and on the other for defraying the costs of the community as such, i.e., for war, religious worship, etc."). Volume 3 of "Capital" discusses this point in more detail, using Polish and Romanian government officials as examples.

"A survival of the old communal ownership of land, which had endured after the transition to independent peasant farming, e.g., in Poland and Rumania, served there as a subterfuge for effecting a transition to the lower forms of grand-rant. A portion of the land belongs to the individual peasant and is tilled independently by him. Another portion is tilled in common and creates a surplus-product, which serves partly to cover community expenses, partly as a reserve in cases of crop failure, etc. These last two parts of the surplus-product, and ultimately the entire surplus-product including the land upon which it has been grown, are more and more usurped by state officials and private individuals, and thus the originally free peasant proprietors, whose obligation to till this land in common is maintained, are transformed into vassals subject either to corvée-labor or rent in kind, while the usurpers of common land are transformed into owners, not only of the usurped common lands, but even the very lands of the peasants themselves."

As we conclude our discussion of the Asian form per se, we would like to draw attention to the fact that, as we have just seen, the "items for acquiring surplus-value" are at first glance a dualistic explanation. In other words, on the one hand, the attribution of surplus-value to the "sole proprietor" is explained as being based on ownership itself, but on the other hand, it is said to be based on the activities carried out within the community, such as communal stockpiling, insurance, war, and rituals, and on the "dominion of lords" derived from these activities. Therefore, it becomes necessary to organize the relationship between these two logics and show how they can be explained in a unified manner.

This problem is also connected to the next problem regarding the layered nature of ownership relationships in the Asian form. The question is how, amidst the multilayered ownership or possession relationships of individuals, communities, and all-embracing unity, why does the all-embracing unity appear as the "sole proprietor?" Alternatively, if there is no all-embracing unity, it would be "the person who embodies the unity of the community," but even if we put it this way, the same question remains. Even though the existence of individuals is mediated by communities and their unions, why does the multilayered nature of ownership relationships not manifest itself in its own form, and ownership by the all-embracing unity comes to the fore? A "major civil engineering project" causes sole ownership, but why does "management by individuals" only result in "propertyless?" In order to unravel this problem, it is necessary to consider the aforementioned activities that take place within a community and the relationships among the individuals involved. Without consideration of this point, the basis for being the "sole proprietor" will not be clear, and therefore the logic of attributing surplus-value to "sole proprietor" on the basis of ownership will not be clear. However, these issues will be discussed later in Section 7. Following the order of description in "Formations," the next section examines various points regarding the particularity of Asian form.

4. Forms other than Asian Form

"Formations" follows the definition of the Asian form as such, that is, its immediate definition, and then characterizes the Roman form and the Germanic form, respectively, in comparison with other forms, including the Asian form. By looking at them, we can understand how Marx recognized Asian form as distinctive compared to other forms. In the following, I will focus my discussion on the points that correspond to the definition of the Asian form.

(A) Forms of Ownership

In the Asian form, there is no private ownership of individuals as distinct from ownership by the community or by "all-embracing unity," and individuals are merely occupiers of allotted land. On the other hand, in the Roman form, there is a separation and conflict between communal property (national property, public land) and the private property of individuals, and in the Germanic form, communal property (hunting areas, pastures, etc.) is merely a supplement to individual property.[70] These characteristics of property form give rise to a class

70 According to Otsuka (1955), private ownership also exists in the form of "heredium" (residential land and garden land) in the Asian form. Furthermore, in the Germanic form, "community land (grazing land and forests)" is also divided into ownership and incorporated into private ownership relationships.

society peculiar to the Asian form, that is, total slavery pairing up with the "all-embracing unity," and the acquisition of surplus-value by the latter.

(B) Forms of Collaboration

In this regard, the Roman form is particularly taken up and discussed in contrast to the Asian form. In the Roman form, the difficulties a community encounters arise from its relations with other communities (i.e. war). Therefore, in this form, war becomes an important joint activity, from which the community becomes a military organization concentrated in cities. On the other hand, in the Roman form, the difficulties faced by humans in relation to nature are not so great. However, it is the latter aspect that is important in the Asian form, where "individual property is utilized only through collective labor" and the construction of "canals, means of transportation, etc." becomes an important joint work. Indeed, war was mentioned as one type of communal work in the definition of the Asian form. However, "in the self-sustaining unity of industry and agriculture on which this form is based, conquest is not so essential a condition as where *landed property, agriculture,* predominate exclusively." (Marx 1858)

Furthermore, regarding the relationship between the Asian form, Roman form, and Germanic form, as mentioned earlier, Marx did not envisage a so-called unilinear development course in a certain region. In other words, it seems that he did not think that the historical development of pre-capitalist societies would follow this order (Section 1 of this chapter). As indicated by their respective names, they seem to have been regarded as types of pre-capitalist societies in geographically separate regions.[71] However, it was

71 As a geographical typology of pre-capitalist societies, Umesao's (1967) taxonomy of the Old World is known as an ecological perspective that emphasizes natural factors. If we compare the arguments of Marx and Umesao, we can see the following correspondence between them. In other words, according to Umesao (1967), the entire Old World, including Asia, Europe, and

thought that there was a logical developmental relationship between the three types in this order. It was thought that although communal ownership and private ownership were in conflict with each other, the latter would gradually grow and surpass the former as the types progressed.

5. Essential Contents of the Community

From the definition of the Asian form in "Formations" that we have seen so far, and the comparison of it with other forms, we can extract the following characteristics as peculiar to the Asian form, which differs from the latter: ① Underdevelopment of private ownership by individuals, ② strong dependence

---

North Africa, can be divided into two categories, of which Western Europe and Japan are the first. The entire continent sandwiched between them is the second region. The first region originally started out as a barbarian people, but it introduced civilization from the second region, and later went through feudalism, absolutism, and bourgeois revolution, and realized capitalistic modernization. In contrast, the second region includes a large arid region that runs diagonally across the continent from northeast to southwest, and although all ancient civilizations originated in this region, they did not develop feudalism. Afterwards, they created huge despotic empires, and many of them became colonies or semi-colonies of the first region countries. Only recently has the second region begun to follow the path of modernization through several stages of revolution. The second region is further divided into the Chinese world, the Indian world, the Russian world, and the Mediterranean/Islamic world. Comparing Umesao's argument with Marx's pre-capitalist community theory, we find that the Asian form corresponds to much of Umesao's second region, and the Germanic form corresponds to the first region of the pre-capitalist stage. Furthermore, the Roman form is thought to correspond to the Mediterranean and Islamic worlds. According to Umesao, although it belongs to region 2, it is in contact with Western Europe and is the nexus of regions 1 and 2, and is a region with a transitional and intermediate character. Umesao (1967) expresses the transitional and intermediate character of the Mediterranean region as follows. "The Greek and Roman civilizations are often seen as the predecessors of Western civilization, but I believe they are different. They would be in the eastern part of the second region and the Western counterpart of ancient China. Modern Italy belongs to the first region, but it is not the successor of the Roman Empire."

of individuals on the community, ③ the isolation and decentralization of the community and its robustness backed by a self-sufficient living area, ④ the importance of civil engineering projects as collaborative work, and ⑤ the probable, but not inevitable, existence of the "all-embracing unity" that binds multiple communities.

If, as this chapter assumes, the essential content of the community is included in the definition of the Asian form, what should we extract as its core? Let us begin our consideration of this point by confirming the purpose of economic activities in the community.

First, for comparison, we will look at capitalist society, which is the opposite of community society. The purpose of people's economic activities in the capitalist society is to maximize the value of various economic categories such as profit, rent, and labor wage, whether in the short or long term. Regarding the purpose of economic activities in a communal society, the following definition by Marx would be appropriate.

"In all these forms (Asiatic, Roman, Germanic forms of ownership—citer), where landed property and agriculture form the basis of the economic order, and consequently the economic object is the production of use-values – i.e., the *reproduction of the individual* in certain definite relationships to his community, of which it forms the basis" (Marx 1858)

In other words, the purpose of economic activities in a communal society is to produce and secure use-values centered on agricultural products of a quality and quantity that guarantees the survival of the individuals who make up the society and reproduction beyond generations. Aristotle (edited around 300BC) also said, "it (wealth—citer) is for use, that is, for the sake of something further:" In this text, he discusses the state of wealth in ancient Greek communal society, with an emphasis on utility. This is a society that differs, at least on the surface, from a capitalist society in which each person aims to maximize economic value. The purpose of a communal society determined the specific nature of production relations and distribution methods within this society. Therefore, the purpose of economic activities of such a community society will be referred to as the "principle of survival."

Furthermore, in a communal society, people who had the ability and will to work were in principle excluded from becoming unemployed and losing the opportunity to earn a living. On the contrary, in a capitalist society, the existence of a certain amount of unemployed people as an industrial reserve army is an essential prerequisite for the existence of this society in order to substantially include workers in this society. (Marx 1867)

However, it seems that Adam Smith, who lived during the Second Enclosure period in England, did not yet have such a clear view of the purpose of production in a capitalist society. For example, "The Wealth of Nations" (Smith 1776) divides society's savings into three parts: ① Portion reserved for immediate consumption, ② "fixed capital" that generates income without circulating, that is, without changing ownership, and ③ "circulating capital" that generates income by circulating and changing hands. He then discusses the purpose of the latter two types of capital as follows: "To maintain and augment the stock which may be reserved for immediate consumption is the sole end and purpose both of the fixed and circulating capitals." In other words, as shown in the phrase "for immediate consumption," there is a use-value perspective mixed in, and there is no dry way of thinking here that sees the purpose of capital as value proliferation. In the first place, the main theme of Books 1 and 2 of "The Wealth of Nations" is the social increase in a country's material wealth in harmony with its population.[72] It is not an increase in value,

72 The "Introduction and Plan of the Work" of The Wealth of Nations states that "this produce, or what is purchased with it, bears a greater or smaller proportion to the number of those who are to consume it, the nation will be better or worse supplied with all the necessaries and conveniences for which it has occasion."

such as today's GDP.

However, the concept of wealth in this work is heavily influenced by the exchange-value perspective in Chapter 3 of Book 2, which deals with productive and unproductive labor, and as a result, the work lacks consistency as a whole. In other words, labor is divided into productive labor, which "adds to the value of the subject upon which it is bestowed," and unproductive labor, which "has no such effect." The labor of manufacturing worker is shown as an example of the former. Examples of the latter include the labor of domestic servants, the labor of public servants engaged in the judiciary and military, and the labor of several professions such as church workers, lawyers, doctors, writers, actors, and clowns. The work goes on to state that productive labor is sustained by capital, and unproductive labor is sustained by rent, profit, and a portion of labor wage, primarily income from the former two. And it states as follows. In the text, the concept of wealth is replaced by one based on exchange-value.

"Every increase or diminution of capital, therefore, naturally tends to increase or diminish the real quantity of industry, the number of productive hands, and consequently the exchangeable value of the annual produce of the land and labor of the country, the real wealth and revenue of all its inhabitants."

It can be said that this fluctuation in the concept of "wealth" reflects the historical position of "The Wealth of Nations," which is a theoretical reproduction of Britain's capitalist society during its primitive accumulation

---

(Smith 1776). The two factors that determine this proportion are (1) the national productivity of labor, and (2) the division ratio between useful and productive labor and non-useful labor. Based on the assumption that the amount of product supplied depends more on the former factor than the latter, Book 1 discusses the causes of increased labor productivity and the issue of distribution of products to various social classes, and Book 2 deals with the question of the capital stock that determines the amount of useful and productive labor.

period.

In any case, community ownership is the form of ownership of the natural production conditions necessary to realize the community's "principle of survival." In the community, production takes place in some cases by direct collaboration. In other cases, natural conditions of production, such as land, are distributed among member families. The community, or the chief person who personally embodies it, distributes the natural conditions of production based on certain rules that enable the survival and reproduction of family members. "Formations" posits the Asian form as the prototype of community property, and then considers the Roman and Germanic forms to be things that are eroded by private property, which is the negation of community. However, on the other hand, "Formations" argues that the same purpose of economic activities as in the Asian form persists in the Roman and Germanic forms.

By the way, the community produces and secures the quality and quantity of use-value necessary for the survival and reproduction of all family members. Is the ownership form of community ownership sufficient to achieve this goal? The answer would be no. This is because while community ownership is a guarantee for the survival and reproduction of its members in the input dimension of the primary means of production, it also requires a guarantee for community members in the output dimension. These two dimensions do not need to be considered separately when production is carried out through direct collaboration. However, when family members receive allocated land, the gap between the two dimensions may become apparent. This will become clear if we recall the following situation. Even if a family receives the necessary and sufficient amount of land from the community, the family may fail in the process of managing such resources, or may be affected by natural disasters, unexpected accidents, or illness. As a result, the family may not produce enough consumer goods or inputs for the following year. It must also be an important function of the community to adjust the consequences of such risks in a broad sense after production so that each of the constituent family units can share the consequences. The method for achieving this is the stockpiling and redistribution of products, which we discussed in detail in Section 3 of this chapter. "Formations," which deals with the theory of forms of ownership, only briefly points out this point in the definition of Asian form, but does not constitute its central point. However, in clarifying the purpose of the community's economic activities and the actual content of activities to realize that purpose, this point, along with the specific method of distributing land, and so forth, can constitute an important point of discussion. In the descriptions of collaborative work in communities seen in Section 4 of this chapter, civil engineering projects and war are emphasized, perhaps to emphasize the contrast between three forms. However, if we consider the above, we can even say that collective labor for stockpiling has a more fundamental meaning. Alternatively, it is necessary to elucidate the nature of material circulation within a community based on the aforementioned Polanyian theory of redistribution and reciprocity. This also means reconsidering "community regulation", focusing on its protection and relief functions. Otsuka (1956b) emphasizes the aspect of oppressing individuals of "community regulation."

However, it would be premature to conclude that pre-capitalist societies were free from poverty and famine just because they had such protection and relief functions (Devereux 1993). It is thought that these were societies that constantly faced general poverty and famine due to their low level of productivity, and that their protection and relief functions were activated for collective defense against this. In pre-capitalist societies, it was rare for specific individuals to suffer from starvation due to the protection and relief functions of the community, but they were societies that were prone to general starvation due to low productivity.

6. Redistribution of Wealth in the Inner Niger Delta

As this chapter focuses on theoretical considerations, the following description is somewhat irregular. In order to provide the reader with a more concrete image of the redistribution of wealth within a community, I would like to summarize the results of a rural survey conducted by the present author at the beginning of this century in the middle reaches of the Niger River, just below the Sahara Desert.[73]

First, a general description of the survey area is given. The total length of the Niger River, which crosses West Africa, is 4,200 km. It is the third longest river on the African continent, after the Nile and Congo Rivers. This great river starts from the Tambi Valley in Guinea at an altitude of 850 meters and heads north, but after passing through Bamako in Mali, passing through Mopti and reaching the vicinity of Timbuktu, it changes direction and heads south, forming a large bend. In the middle reaches of the Niger River around this large bend, an inland delta of approximately 3 million ha has been formed from Macina to Timbuktu (L'Afrique Authentique 2003). (**Figure 4-1**)

The annual precipitation in this delta is low, ranging from 200 to 700 mm, and the climate zone is the Sahel. There, the seasons are clearly divided into the rainy season and the dry season, with annual rainfall concentrated from July to September, but no rain at all from November to February. Temperatures usually exceed 40 ℃ during the day even in the shade in April and May, before the rainy season, but they drop slightly with the arrival of the rainy season. If people were to farm in this area using only rainwater, they would only be able to grow pearl millet, fonio, and sorghum. However, most of the delta, which is covered with flood water from the end of July to mid-November, becomes paddy fields. In fact, this area is the birthplace of many

73 For details of the survey results, see Yamazaki (2007, Collected Works Vol. 4).

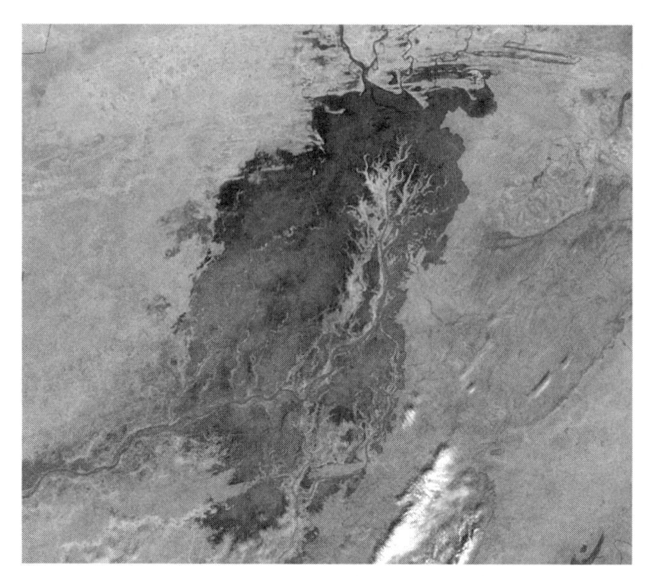

Figure4-1：Inner Niger Delta (NASA-2007)

types of wetland grasses (Nakao 1969), and is one of the world's two major birthplaces of rice cultivation, along with southern China and its surrounding areas. The history of rice cultivation in the Delta can be traced back 3500 years, according to Portères (1950). In addition, due to the rich potential of rice production that still exists today, France, a former colonial power, has been conducting extensive research on rice cultivation in this area. Valuable information has been passed on to Japan through research conducted in anthropology and agronomy.[74] I visited this area three times from 2003 to 2004

74 The following is a list of Japanese literature that introduces the current state of rice cultivation in the Inner Niger Delta and its surrounding areas, although it cannot be called a comprehensive introduction. a) Kawata (1981) describes traditional rice cultivation in this region. b) Takezawa (1984) introduces the actual situation of rice cultivation around Dia, Diafarabe, and Gao. c) Japan Association for International Collaboration of Agriculture and Forestry (1986) provides an overview of Mali's irrigated agriculture, rice cultivation techniques and farmers' lives under the natural submergence system, and the development process of the Office du Niger up to the mid-1980s. d) Cho et al. (1979), e)

with the cooperation of France's CIRAD (Centre for International Research in Agriculture and Development). I stayed there for a total of three months and conducted research activities targeting rice farmers and related organizations. In 2006, a supplementary survey was conducted over a two-week period.

The farm survey, some of the results of which are introduced in this section, was conducted in Ngimitongo Village, Socoura District, Mopti Prefecture.[75] The village is located on a slight elevation above the flood plain, and according to the 1998 census, it has 679 residents. Looking at past census data, the village's population increased by 51% from 1976 to 1987, but only increased by 3% from 1987 to 1998. The period when the village's population increased significantly coincided with the introduction of highly artificial irrigation system[76] for part of wet rice cultivation. It is suggested that the increase and stabilization of food supply through rice cultivation using this system resulted in an increase in the village population. Mopti, the town where the Bani River meets the Niger River, is located at 15 degrees north latitude and 4 degrees east longitude. It is a trading city on the southern tip of the Sahara with a population of approximately 100,000, and is also the center of local government.[77] Ngimitongo is located approximately 10 km northward,

---

Hamamura et al. (1992), f) Hirose et al. (1997), the above three documents briefly introduce the existing irrigation systems in the Inner Niger Delta. g) Oji (1993) points out that there is regional diversity in the location environment of "rice fields" in natural submergence system. h) Oji (Kawata 1999) compares the "rice cultivation landscape" of the Inner Niger Delta with that of monsoon Asia.

75 Mali has an administrative structure in which administrative units such as districts, prefectures, and states are built up in stages on top of the natural villages that existed before the colonial period. The surveyed village corresponds to the level of natural villages.

76 Yamazaki (2007) introduces this system as the "*petit périmètre irrigué villagois: PPIV.*"

77 Mopti is "a cosmopolitan town. It is a melting pot where different ethnic groups (Fulbe, Bozo, Songhai, Bambara, Dogon) coexist in perfect harmony, and it is

downstream, along the Niger River from Mopti. From Ngimitongo to Mopti, there is a gravel road on the river embankment that can be accessed by car, motorbike, bicycle, cart, and foot. This river embankment was constructed in the 1970s as part of a partial control system. Before that, it was virtually impossible to travel by land between Mopti and Ngimitongo during the peak of the rainy season. In addition to this land route, it is also possible to travel by regular-run canoe with an outboard motor using the river. However, farmers generally prefer land transportation over river transportation due to its lower fares and ease of transportation, except in cases where large quantities of goods need to be transported. Ngimitongo is located in a location that can be described as a suburban area due to its proximity to Mopti and its easy access by land. However, in Ngimitongo, research has revealed almost no development of commercial agriculture or commuting to work that takes advantage of this location. This shows that the economy of the target area is still based on a subsistence lifestyle.

The most important crop in the village is rice. The main methods of cultivation there are extensive rice cultivation using natural submergence and partial control methods, using floodwaters from the Niger River and floating rice.[78] In addition, wet rice cultivation using the aforementioned highly artificial irrigation and improved varieties is practiced in some areas. The total cultivated area of the 31 target farmers[79] was 186 ha,[80] based on the figures obtained from individual farmers, of which most (98%) was paddy fields.

---

also a major center for crafts and tourism." (Kawata 1997) The population of Mopti according to the 1998 census was 80,872.

78 For information on these rice cultivation methods, see Yamazaki (2007).

79 Regarding the method of randomly selecting target farmers, see Yamazaki (2007).

80 In the target area, as part of a development aid project, the area of cultivated land of individual farmers was actually measured using ropes (according to Mr. Sangare, a research collaborator).

Crops other than rice include kidney beans in the fields and shallots near the residential areas. Rice is the only agricultural product that is commercialized, but only 25% of its production is sold. Furthermore, this ratio is high in Ngimitongo because the payment for the use of paddy fields for the artificial irrigation system is made in cash. For example, in another nearby village (Kamaka) where cash payments are not required, this ratio was only 3%. Agriculture in the target area and its surrounding areas has an extremely strong self-sufficiency character.

Housewives were responsible for milling the rice by hand using wooden mortars and vertical pestles, transporting the rice to Mopti markets using two-wheeled carts pulled by donkeys, and selling the rice at street stalls. The average weight of paddy that one adult woman mills per day is said to be 20 to 25 kg.[81] There are several markets in the vicinity, but each market is open only one day a week, and on that day, business is open from around 10 a.m. until just before sunset.

Other sources of cash income other than rice sales are livestock sales, remittances from migrant workers, and incomes from herdsman's works and other off-farm activities carried out within the village. Of these, the other off-farm activities that were discovered through the survey were fishing, retailing of miscellaneous goods and food within the village, transportation by two-wheeled donkeys, weaving, bicycle repair, and livestock trading, but the variety is not abundant. Women smoke the fish caught, sell the amount that exceeds their own consumption at the market, or barter it for paddy. What is striking is that the Bozo people, who have been a group of professional fishermen for the past several hundred years, are all engaged in fishing. Unlike in rural Southeast Asia, we do not see a situation where various business activities that can be called "miscellaneous jobs" are developed in a

81 Based on an interview at Kobaka (November 19, 2003).

multilayered manner within the village. Furthermore, among the non-agricultural activities in the study area, weaving is characterized by hereditary occupation and endogamy within the occupational group, and is based on a caste-like system. In West Africa, in addition to weaving, other caste-based occupations include woodworkers, storytellers/musicians (griots), blacksmiths, pottery makers, leather craftsmen, and gold and silversmiths. (Ogawa 1987, Kawata 1997, Takezawa 2015)

Farmers purchase livestock if they have extra cash. The following is generally said to be the main reason why farmers own livestock. The purpose of livestock ownership is to compensate for fixed tax payments and for grain shortages, under unstable climatic conditions and resulting fluctuations in harvests. (Kawata 1997)

The reasons for cash demand by target farmers include ① tax payments and ② grain purchases during poor harvests. ③ In addition, as mentioned earlier, payment of fees for use of paddy fields for artificial irrigation systems is a characteristic feature of this village. Moreover, cash is required to purchase the following items; ④ agricultural tools such as sickles, plows, and hand hoes; ⑤ daily goods such as earthenware, clothing, and mosquito nets; ⑥ consumables such as condiments, kerosene, tobacco, and medicine. There are also other expenses associated with production activities, such as ⑦ allowances for herdsmen when owning livestock, and fees for nets and fishing hooks when engaged in fishing. In addition, ⑧ household goods such as TVs, radios, bicycles, and motorcycles, and ⑨ means of transportation such as canoes and motorcycles may be purchased with cash.

The Inner Niger Delta region is a place where many ethnic groups (also called tribes) coexist. The number of ethnic groups in the region depends on how they are counted, but for example, Sanogo (Kawata 1997) classifies the residents into six groups; Fulbe, Bozo, Bambara (Banaman), Soninke, Songhai, and Bobo. In Ngimitongo, ethnic diversity can be seen even within the village. But the dominant one is the Fulbe (French: Peul). Of the 31 households surveyed, 18 (58%) were Fulbe, 5 (16%) were Bozo, and 8 were other. The Fulbe are an ethnic group that has traditionally made a living from livestock farming, but some groups settled down at some point and began to engage in farming. Fulbe in the target area is an example of the latter. Bozo's main occupation is fishery.[82] The Bozo is said to be the earliest ethnic group to settle in the Niger River basin. (Griaule 1948) Although the population is ethnically diverse, they are all Muslims. And this is true for Mali as a whole.

Let us end this rather long introduction and get to the main topic of this section. What follows is to show that there were considerable differences among farmers in the amount of paddy that farmers kept in their homes without selling it after harvest in the survey year. After that, think about what it means. The uses of paddy that is not commercialized are as home-cooked rice, goods for barter within the village,[83] seed for the next year, tribute to the village chief, the "lord of the land," or even the imam,[84] and charity to those in need. If we divide the total amount of paddy held by all target farmers (105 tons) by the total number of family members (230 people), after excluding the amount allocated to seed paddy, the average amount of paddy held by each family member is calculated. It was calculated to be 0.457 tons. Dividing the total amount by the number of farms and calculating the average amount per farm household was 3.4 tons. However, as shown in **Figure 4-2**, there was

---

82 There is a series of studies by Takezawa on the society, economy, and religion of Bozo. (Kawata 1997, 1999, Takezawa 1988, 2008, 2015)

83 Takezawa (2008) states that the barter weight ratio of paddy and fish in the delta was 2:1.

84 Imam means "guide." In Sunni usage, this term usually refers to a person who stands in front of the congregation during Friday corporate prayers and leads the rituals of worship. (Izutsu 1981) In the target area, he was called marabout, and played various roles related to village life, such as Quran school teacher.

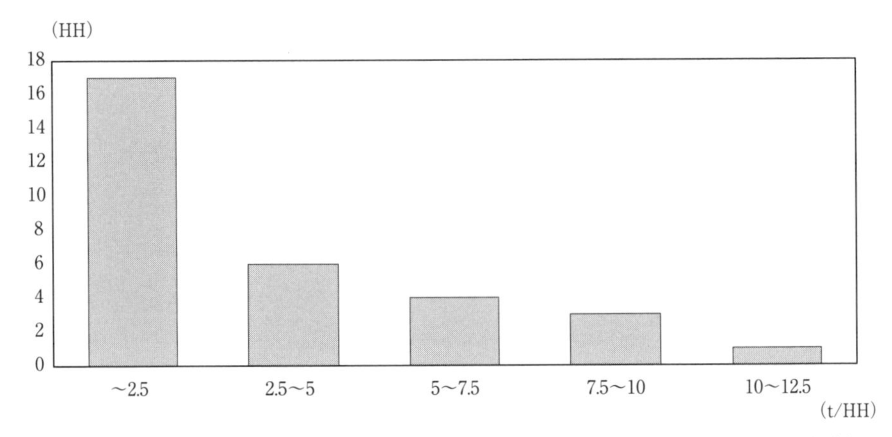

Figure4-2 : Number of households by self-holding amount of paddy per household (Ngimitongo)

Note 1) Amount of paddy held at home = Paddy production amount - Paddy sales amount - Seed amount
(Source) Created from an interview survey conducted in 2003-04.

considerable variation in this amount among farm households. There was a tendency for "upper farm households" with larger farms to keep a larger amount of paddy in their own homes.[85] Also, when comparing the amount per family member, there are large disparities between farm households (**Figure 4-3**). There was a 38-fold difference between the lowest (0.068 t/person) and highest (2.553 t/person) farm households. Here, too, it was observed that the larger the "upper farm households" are, the greater the amount of paddy they

85 The simple regression coefficient between the equivalent amount of paddy fields (x), which is an indicator of the farm's management scale, and the amount of paddy held by the farm household (y) was significant at 1% on one side ($y=0.13x-0.49$, t value of regression coefficient=9.02, n=31). In order to calculate the equivalent amount of paddy fields, first convert the area of paddy fields with a relatively high level of maintenance and fields to the area of "paddy fields" that are naturally flooded with the lowest level of maintenance, using their yield as the standard. Furthermore, these three items are aggregated per farm household. The equivalent amount of paddy fields indicates the scale of management. For details on equivalent amount of paddy fields, please refer to Yamazaki (2007). Also, the reasons why there are differences in the scale of business among farmers are explained in the same book.

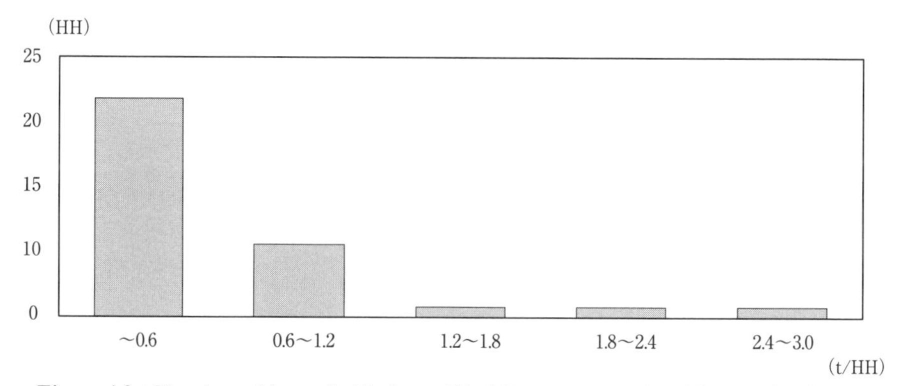

Figure4-3 : Number of households by self-holding amount of paddy per family member (Ngimitongo)

Note 1) Amount of paddy held at home = Paddy production amount - Paddy sales amount - Seed amount
(Source) Created from an interview survey conducted in 2003-04.

keep in their own homes per family member.[86]

Next, what kind of interpretation should we give to the fact that there is such variation? The 38-fold difference between farm households in the amount of paddy they keep per family member is too large to be explained solely by differences among farm households in the amount of food consumed by family members. Rather, we need to focus on the mechanism that exists in the target area to absorb surplus rice from farm households. First of all, Zakat (ordained charity) is a part of Sharia (Islamic Law), [87] along with Shahada (a profession of faith that proclaims that there is no god but Allah and that Muhammad is His Messenger), prayer, fasting in Ramadan, and the pilgrimage to Mecca, is one of the formalized practical obligations (the five pillars of Islam). [88] (Chuo

---

86 The simple regression coefficient between the equivalent amount of paddy fields (x) and the amount of self-held paddy per person (y) is still significant at 1% on one side ($y = 0.02x + 0.09$, t value of regression coefficient = 3.13, n = 31).

87 Regarding Islamic Law, see Izutsu (1981).

88 Mauss (1925) states that the Arabic word "sadaqah" originally meant justice. Furthermore, he sees the origin of the idea of charity in the moral ideas regarding gifts and property, and the idea of sacrifice. Regarding the amount of charity one should give, the Quran contains the following statement: "They ask

University/Japan Institute of Comparative Law 1978). In Ngimitongo, a devout Muslim village, these duties are faithfully practiced, with the exception of the pilgrimage to Mecca, which requires financial reservations.[89] For example, the head of a household that harvests about 1.5 tons or more of paddy per year is said to offer one-tenth of the harvest as zakat. The surplus absorbed from farm households in this way is sometimes passed through tribute to the "lord of the land," the village chief, or the imam, and a portion of it is consumed there, and sometimes it is appropriated for reserves in times of crop failure. However, a significant portion of the surplus is used as handouts among farmers, as donation from the haves to the have-nots. In other words, there are cases in which paddy is transferred directly from the "upper layer" to the "lower layer" between farm households without going through the village authority. However, according to Sahlins (1972), this provision of livelihood aid from the haves to the have-nots is common in primitive societies where a significant percentage of households periodically fail to earn a normal living. In primitive societies, therefore, individuals are generally threatened with starvation only when the society as a whole is in the same predicament. (Polanyi 1944) The "primitive chief" who is the personification of the "public economic principles" that constitute these contents is, so to speak, a waterway through which wealth flows. The reason why wealth is once concentrated in his place is so

you what they should donate. Say, whatever you can spare." On the other hand, the meaning of "almsgiving" in Buddhism to help the poor seems to be weak. In Suttanipāda, a collection of Buddha's discourses, in response to a young Magha question about who are "people worthy of receiving alms," the Buddha says, "Indeed, one should walk in the world without attachment, and be without attachment." Although the answer points to practitioners with certain inner qualities such as "a perfect person who is penniless and has mastered himself," he does not mention the poor, and in the first place, the economic condition is not a standard for giving "alms."

89 Even in medieval Japan, there existed a virtuous ideology that "the rich must give charity." (Sakurai 2011)

that it can flow freely from there again. (Polanyi 1944) Or much wealth passes through his hands, but he never owns it. (Lévi-Strauss 1955) Furthermore, the "public economic principles," which have an extra-economic nature, are probably expressed subjectively by community members in the target area as Islamic moral norms. (Otsuka 1955)

From the above, within the community of *tendresse*, paddy is transferred between farm households from the "upper layer" to the "lower layer" without going through the market, sometimes via the chief. The disparity between farm households in the amount of paddy sold, which would have been even greater if this had not been done, has been offset, even if only partially, through such obligations and almsgiving practices.[90] If we focus on this type of movement of paddy within villages without going through the market, we can see that the self-sufficient character of rice cultivation is not completed at the level of individual farm households, but is realized within the spread of the village.[91]

7. Conclusion—Community and Class Society

As we saw in Section 3, the portion of products that is used for stockpiling and redistribution is the initial form of surplus-value. Therefore, it is also the

90 The sharing of food among local and blood relatives, and the resulting "leveling of consumption" among them, seems to be a widely observed social phenomenon in Africa even today. Regarding this point, see Sugimura (2004).

91 Hyden (1986), while introducing the research results on Guinea-Bissau in the mid-1980s, states that when farmers produce surplus agricultural products, the order of priority for how to dispose of them are; (1) giving alms to the poor, ② donations, ③ barter, and ④ sales to merchants. Furthermore, Magasa (1978) has the following description. "What authors call 'parasitic spending' is the concrete extraction that takes place in reality under traditional structures. It weighs heavily on the shoulders of all working people in this part of Africa. This is done through social solidarity (underwriting by working people of expenses for 'unproductive' members—young people, old people, women, sick people—) and moral compulsion."

material basis for the second-order formation of dominance-subordination relationships based on the community as a result of its transformation. Therefore, the mechanism by which a community, established for the purpose of individual survival, transforms into a class society can be elucidated by focusing on the specific ways in which wealth is stored and redistributed in the community. Furthermore, in addition to the functions of stockpiling and redistribution that have been emphasized in this chapter, there are also "common activities" (Marx 1894) or "social functions" (Engels 1878) that are universally found in communities, such as judicial, police, military, water conservancy, and religious functions. In return for carrying out these tasks, and under the political control and "ownership differentiation" (Weber 1924) of specific groups and classes that are established as a result of these tasks,[92] some or all of the surplus-value produced within the community is transferred to these groups and classes.[93] Alternatively, there are inherent communal practices and ideologies that create social nodes (like the chief we saw in the previous section) where surplus-value is concentrated for stockpiling and redistribution. Individuals who take initiative become fixed as a class through hereditary inheritance while also performing other social functions. As a result, the social functions of society have become more independent, and it can be

92 "An even more significant change was that the *mitegra*, the sacred tree, gradually created a special class of people who held it. In the past, within a clan, kin, or party, the person who would take on that role would have been determined naturally, but this depended on the person's status and circumstances, so it was like a manifestation of divine will. At first, simply having this gift in hand was what made a person special. Later on, the relationship became reversed, with people using this tree to serve at festivals because they were special people." (Yanagida 1942)

93 According to Montesquieu (1748), the lord's jurisdiction under the French feudal system was nothing more than the right to demand the payment of atonement amounts and the right to demand fines as established by law. It was a right inherent in the fiefdom itself, and a profit-making right that formed part of it. From this arose the principle that jurisdiction is patrimonial.

said that a secondary stratification has been formed.[94] (Marx 1881) However, among these various "common activities" and "social functions," the functions of storing and redistributing wealth are thought to have had the most fundamental significance in the establishment of class society. The first reason is that the original purpose of community economic activities was based on the "principle of survival." Secondly, and more importantly, they were the very points where surplus-value separated from the part of the product that corresponded to the other means of subsistence.

An example of this can be seen in Lefebvre's description (1939) of the period before the French Revolution. It is stated that a large portion of the tithe, which was originally collected for the purpose of helping the poor, benefited the high-ranking clergy and feudal lords at the time.[95]

"If the money collected as tithes had been used for the maintenance of festivals, the upkeep of village cathedrals and priest's houses, and especially for the relief of the poor, the burden would not have been abhorred. In reality, however, the increase in tithes often benefits prelates such as bishops, abbeys, and cathedral chapter. The tithes are 'conferred' and even became the property of a secular lord. In contrast, the village priest's share was no more than a small tithe." [96]

However, once the surplus-value was concentrated in the ruling class, as

---

94 From the charismatic community where people lived in a communist manner based on gifts, charity, and booty, a class of rulers' assistants arose who made a living from usufructs of land, profit, or wages in kind or wages, in short, by fiefs. Now, the legitimacy of power has come to be derived from things like bestowing and appointing. And that meant, in most cases, the *patrimonialization* of the ruler's power. (Weber 1920-1921)

95 It was Charles the Great (742-814) who established the tithe system. (Montesquieu 1748)

96 A large tithe tax is levied on the four major grains (wheat, rye, barley, and oat). A small tithe tax is levied on other grains, vegetables, and fruits. A tithe was also sometimes levied on livestock products.

long as the communal aid system was functioning, at least a portion of it was appropriated for stockpiling and redistribution for the people. This is because the ruling class could assert and secure the legitimacy of its own power only as the embodiment of communal functions. (Ouchi 1977) However, a large portion of surplus-value was still appropriated for the expenditure of the ruling class. In some cases, poverty relief fees could also be collected separately. Therefore, the concentration of surplus-value in the hands of the ruling class also triggered the decline of the people and the concentration of land. In that case, while there were groups of people who lost their land and became debt slaves or became proletarians, this land was concentrated on the side of the ruling class. However, there was a possibility that this concentration of land would come to a halt as long as communities continued to function. This is because, on the one hand, there were cases in which sales of community's land were prohibited in the first place, and on the other hand, as an extension of this, there were cases in which the results of land sales were periodically invalidated. Let us once again look for an example of the latter in the Old Testament.

The book of Leviticus states that in the year of Jubilee, which is proclaimed every 50 years with a trumpet, the entire population will be freed from debt slavery, and the land sold will be returned to its original owners. (25 no. 9, 10, 13) This is because the land belongs to God, and the people are only strangers and live as foreigners in God's land. (25 no. 23) However, regarding the price of land for sale, Leviticus states the following: "If you sell anything to your neighbor, or buy from your neighbor, you shall not wrong one another. According to the number of years after the Jubilee you shall buy from your neighbor. According to the number of years of the crops he shall sell to you. (25 no. 14,15)" In addition, the right to buy back sold land is given priority to the person who sold it or his relatives, but the price of the buyback is calculated according to the number of years remaining until the next Jubilee. (25 no. 23-27) Judging from this description, what is referred to here as "land sales" should today be regarded as a paid lease of land whose period is divided by the Jubilee year. However, Weber (1920) argues that the year of Jubilee was a theological concept during the Babylonian captivity (sixth century B.C.) and was never implemented. On the other hand, he posits a connection between this concept and the laws actually applied in ancient Israel (practice of freeing debt slaves during wartime and giving priority rights to relatives to purchase the land of ruined persons).

Furthermore, according to the following description in "The Wealth of Nations," even after the establishment of large landowners with the right to judge and conscript troops, in pre-capitalist societies large landowners continued to exercise generous treat over the people of their territories. In addition, a system of mutual aid of the community was functioning through charity and entertainment provided by the clergy.[97]

"In a country which has neither foreign commerce, nor any of the finer manufactures, a great proprietor, having nothing for which he can exchange

97 Caesar (ca. 52-51 BC) wrote the following about the customer service customs of the Germanic people at the time. "They defend from wrong those who have come to them for any purpose whatever, and esteem them inviolable; to them the houses of all are open and maintenance is freely supplied." Tacitus (AD 98) also has a similar statement: "No nation indulges more profusely in entertainments and hospitality. To exclude any human being from their roof is thought impious; every German, according to his means, receives his guest with a well-furnished table. When his supplies are exhausted, he who was but now the host becomes the guide and companion to further hospitality, and without invitation they go to the next house. It matters not; they are entertained with like cordiality. No one distinguishes between an acquaintance and a stranger, as regards the rights of hospitality." Montesquieu (1748) quoted both of them and lamented that "such hospitality is extremely rare in commercial countries." In a scene from "Hernani "(1830), Hugo's revenge drama set in Spain around 1520, the lord of Saragossa welcomes Hernani, a wandering bandit who disguises himself as a pilgrim and beggar, as a guest.

the greater part of the produce of his lands which is over and above the maintenance of the cultivators, consumes the whole in rustic hospitality at home." "Before the extension of commerce and manufacture in Europe, the hospitality of the rich, and the great, from the sovereign down to the smallest baron, exceeded everything which in the present times we can easily form a notion of." The largesse of the large landowner is due to the fact that "the capacity of his stomach bears no proportion to the immensity of his desires, and will receive no more than that of the meanest peasant." (Smith 1759) However, as Mauss (1925) states, this unidirectional gifting reproduces, on a moral level, a relationship of domination and subordination between the giver and the receiver.

"To give is to show one's superiority, to be more, to be higher in rank, magister. To accept without giving in return, or without giving more back, is to become client and servant, to become small, to fall lower (minister)."

Turning now to medieval Japan, according to Ishimoda (1946), during the Kamakura period, local feudal lords, as landowners, had the following various rights over farmers; the right to impose servitude, jurisdictional power, police power. These were inextricably linked to the various duties that feudal lords had towards their peasants.

"The rights of the feudal lord could not exist apart from the collective living structure of the village society organized under the feudal lord. The feudal lord had a wide range of duties towards the peasants. These included the protection and relief of farmers in times of crisis, the maintenance of shrines and temples that were the center of community life in villages, and various daily cares. Sovereignty could continue to exist through the fulfillment of these obligations. The rights of feudal lord were not absolute, but were limited in many ways by the traditions and customs of a long group life."

However, this mutual assistance system was undermined with the development of capitalist systems in "commerce and industry." As long as the object of desire was limited to use-value centered on the means of livelihood, there was a natural limit to the total amount of desire of large landowners. However, with the development of the commodity economy, the scope of luxury goods expanded, and as exchange-value came to be emphasized, it could become an infinite quantity, and in fact, infinity became conscious.

"(Foreign commerce and manufactures) gradually furnished the great proprietors with something for which they could exchange the whole surplus produce of their lands, and which they could consume themselves without sharing it either with tenants or retainers." "As soon as they could find a method of consuming the whole value of their rents themselves, they had no disposition to share them with any other persons." (Smith 1776)

"The gradual improvements of arts, manufactures, and commerce, the same causes which destroyed the power of the great barons, destroyed in the same manner, through the greater part of Europe, the whole temporal power of the clergy. In the produce of arts, manufactures, and commerce, the clergy, like the great barons, found something for which they could exchange their rude produce, and thereby discovered the means of spending their whole revenues upon their own persons, without giving any considerable share of them to other people. Their charity became gradually less extensive, their hospitality less liberal or less profuse. Their retainers became consequently less numerous, and by degrees dwindled away altogether." (*ibid.*)

Yasusada Miyazaki's "Encyclopedia of Agriculture" (1697) is said to be a microcosm of agriculture in western Japan from the middle of the early Edo period to the beginning of the mid-Edo period, and geographically centered on Kyushu. The following description shows how the once-strongly functioning system of protection and relief for farmers was shaken as the commodity economy spread.

"In ancient times, if people were frugal and did not spend lavishly, they would be able to accumulate wealth, not only in the great capitals of the world, but also in the capitals of other countries. Each storehouse was full, as proper governance was in place to alleviate disasters and provide relief to the poor. In addition, even villages large and small in the provinces built storehouses for emergency supplies of rice, saving people from misfortunes such as fires and diseases. All of these are related to political matters and can be found in old texts. As the king of the world, he knew well that he should be like the wise kings of ancient times, who would bless all people on behalf of heaven, and make their duties as parents of the people. However, in the later generations, even the humble common people forgot the manners of the ancient times and loved luxury and did not know anything about extravagant. Is this a sign of this era?" "Now, if they suffer from a bad harvest, they will spend all their household possessions, put their fields in pawning, provide a lot of interest to the rich merchants, and suddenly become destitute. They only grieve for the time and the land. It's not just the fault of the time. For example, one out of three years is a good year, one year is a bad year, and the remaining year is somewhere in between. If this is the case, if people save the surplus from good years and use it to cover the shortfall from bad years, there will be no serious shortage."

By the way, speaking about the role of religion in class society, in addition to the "ideology that justifies the appropriation of the surplus generated within the community to stockpiles" mentioned in Section 3, it also included what Weber (1920-1921) called the "theodicy of happiness," which we will see next, was newly added and began to contribute to the justification of domination.

"By holding that suffering was a sign of God's hatred or a sign of hidden transgression, religion catered to a very wide range of psychological needs. Happy people are not satisfied with just the fact that they are happy. More than that, they begin to demand justification for their own happiness." "This justification of happiness is what religion must serve for the external and internal interests of all rulers, wealthy people, victors, and healthy people, that is, happy people. This justification is the most common formula for the work of religion. This is called the theodicy of happiness."

Furthermore, this "theodicy of happiness" corresponds to a sense of "sin" on the part of the people. As an extension of this, faith in the "Savior" who will take on the burden of people's "sins" emerges.

"To determine what is responsible for the affliction, that is, to make confess 'sin,' and then to advise by what action the affliction may be removed. This has now become the typical work of sorcerers and priests." "The development of the 'Savior' faith, prompted by a typical and recurring state of poverty, meant a step forward on this trajectory." "Messianic faith, announced by prophecy, has in most cases continued to take root, especially among the disadvantaged social strata, and has completely replaced witchcraft among them. Or perhaps it became a rational supplement to magic."

However, "there were too many 'unfounded' worries for people. And too often, not only according to 'slave morality' but also according to the standards of the ruling class, it was not the best people but the 'bad ones' who were most successful."

In medieval European moral philosophy, the possession of virtue and happiness were considered almost incompatible.[98] (Smith 1776) In order for trust in God to continue without contradiction to the "impossibility" inherent in the "theodicy of happiness" (the discrepancy between achievements and destiny

98 The break with medieval moral philosophy is said to have been an important opportunity for Adam Smith to form economics (Mizuta 2001). However, this break was not made by Smith alone, but rather had a long run-up through the Scottish Enlightenment. Regarding the Scottish Enlightenment Movement, see Sakurai (2009).

seen in the world), the following points were explained as reasons for suffering and injustice. (Weber 1920-1921) ① Sinful acts committed by individuals in their previous lives (in the case of reincarnation of souls). ② The sins of the ancestors that affect the third and fourth generations. ③ The very corruption of all creation (in the most principled form). In other cases, the cause of evil was not God but rather demons and evil spirits, and in these cases magic was organized. (Weber 1920) The promise to those who keep the commandments is the expectation that individuals will be able to live a better life in this same world in the next life (reincarnation of souls), the expectation that their descendants will be able to live such a life (the kingdom of the Messiah), and the expectation of a better life on the other side of the world (paradise). (Weber 1920-1921)

For example, in the Islamic holy book, the "Quran," the following view of the future is expressed repeatedly: Those who fear Allah and be modest, believe in His omens (existence of the "Quran" is one of them), perform the duties of worship well without worshiping other gods, give generously in charity, and fight holy wars against the evil believers. These people will be rewarded with eternal passage to paradise on the day of the resurrection of the dead, which will surely occur at some point in the future. On the other hand, those who neglect these good deeds at the devil's instigation will go to hell and suffer there forever. It is said that most of the original followers of Islam, founded by Muhammad, were the poorest people. (Izutsu 1958)

Chapter 5 : Community and National Differences of Wages

1. Theme

In recent years, Japanese companies have come to rely on several sources to provide the labor necessary for their capital accumulation. (Yamazaki 2014a) One such area is the non-capitalist sector, centered on agriculture in developing countries. On the one hand, Japanese companies are expanding overseas and hiring local labor force in the countries they are expanding into. On the other hand, the introduction and employment of foreign labor into the country is progressing, albeit at a slower pace than in the advanced capitalist economies of Europe and North America.

The task of this chapter is to focus on the national disparities in labor wages (particularly the relative low wages in developing countries) and examine the factors behind these social phenomena. However, what I will do here is not to assume and analyze a concrete situation, such as Japan and the developing countries in which Japanese companies have invested (or developing countries that send labor to Japan) in recent years.[99] Instead, I deal with the problem of wage disparities that exist in a more general way, namely between developed countries in general and developing countries in general. Therefore, I limit the task to organizing the theoretical points necessary to consider this problem.

By the way, when setting a topic like this, the first thing that must be taken into account is the general discussion regarding the determination of wage levels, which will be outlined below, that is, wage theory.

According to wage theory, wage is the price of the commodity called labor-power.

It is directly determined by the supply and demand relationship for labor-

99 For more on this issue, see Collected Works Vol. 4 (2021) Part 1.

power, which fluctuates with business cycles and capital accumulation, and the value of labor-power is at the center of the fluctuations. The value of labor-power is determined by the total value (price)[100] of daily necessities with the quality and quantity, which are necessary on a social average for the reproduction of labor-power. Furthermore, the total amount of value is determined by how it is earned, namely divided among family members. Furthermore, the total value (price) of daily necessities is further broken down into two components; (1) the range of daily necessities that determines the value of labor-power (i.e. standard of living issue), and (2) the value (price) of those necessities (i.e. price level issue). In addition, when considering the problem of supply and demand in the labor market, we must take into account the pressure of surplus-population. It functions as a weight that binds wage to the level of the value of labor-power despite the constant accumulation of capital.

Next, there is also the question of how non-capitalist modes of production are related to this way of determining wage. For one thing, given the level of labor reproduction cost, to what extent is the burden borne by the capitalist production sector, and on the other hand, to what extent is the burden borne by other non-capitalist modes of production? In other words, there is a conceptual separation between the reproduction cost of labor-power, which can be partially borne by non-capitalist modes of production, and the value of labor-power, which should be at the center of wage fluctuations.

However, the relationship of the non-capitalist modes of production to the wage level is not limited to its role as a partial bearer of the reproduction cost of labor-power. In addition, this relationship may extend to the dimension of

100 Such expression, often used in this chapter, means that price corresponds to value, that is, that price is merely a monetary expression of value. However, today this point requires further consideration based on monopoly price theory.

labor-power pricing theory, as the non-capitalist modes of production become a surplus-population pool and determine the nature of supply pressure in the labor market. Also, when dealing with the problem of non-capitalist modes of production in developing countries, one must take into account the fact that direct producers are sometimes protected by strong communities. What are the unique problems in this case compared to other cases? However, although I am concerned with the non-capitalist modes of production, I do not intend to go back to the feudal mode of production here. What I am talking about here is the non-capitalist modes of production that are "articulated" with the capitalist mode of production and supply labor to it, or at least have the potential to function as such a supply source. Therefore, they can be recognized as concrete forms of existence in independent self-employed farmers and small-scale tenant farmers, who are freed from feudal relationships of personal servitude.

By the way, in Chapter 20 of Volume 1 of Capital, "National Differences of Wages," there is the following classic description of the international comparison of wages. This text explains the factors that determine changes in the amount of the national value of labor-power.

"In the comparison of the wages in different nations, we must therefore take into account all the factors that determine changes in the amount of the value of labor-power; the price and the extent of the prime necessaries of life as naturally and historically developed, the cost of training the laborers, the part played by the labor of women and children, the productiveness of labor, its extensive and intensive magnitude."

My problem with the above-mentioned "value of labor-power" (prior to the conceptual separation between it and the cost of reproduction of labor-power) corresponds to what Marx calls "the price and the extent of the prime necessaries of life, the cost of training the laborers, and the part played by the labor of women and children." However, more specifically, the first two of these three factors are related to the "total value (price) of daily necessities necessary for the reproduction of labor force." On the other hand, the final factor relates to "the situation of how the total value is earned, namely divided among family members."

However, my above-mentioned issue of "supply and demand in the labor market" is not included in Marx's text. This issue is developed as a relative surplus-population in Volume 1 of "Capital," in Part 7, "The Accumulation of Capital," which follows Chapter 20. Therefore, at this stage of the narrative, where this concept has not yet been discussed, it is natural that Marx does not include this issue of "supply and demand in the labor market" among the factors that should be taken into consideration when making international comparisons of labor wages. However, if we take into account the descriptions in Part 7, it goes without saying that this factor must also be taken into account when comparing international wages. Furthermore, Marx does not even touch on the issue of "non-capitalist modes of production" in this quote. It may be possible to see here the methodological orientation of economic principles, which presupposes the total control of capitalist production in society and describes its internal logic. However, as we saw earlier, it is necessary to take into account the "non-capitalist modes of production" as an important factor in international comparisons of labor wages.

On the other hand, in this text, Marx also cites "the productiveness of labor, its extensive and intensive magnitude" as factors to be considered in international comparisons of labor wages. Of these, the "extensive magnitude of labor" is immediately followed by the following statement: "Even the most superficial comparison requires the reduction first of the average day-wage for the same trades, in different countries, to a uniform working-day." In short, this point can be dealt with relatively easily in theory by "adjusting daily wages to equal hourly wages." However, the national differences in labor productivity and labor intensity (intrinsic size of labor) that emerge after time-wages are replaced by piece-wages give rise to complex issues that will be discussed in detail in the next section. These effects are related to the so-called "modification of the law of value," which ultimately results in national differences in price levels.

Among the several characteristics that can be seen in the problem setting of Marx's theory of national disparities in wages that we have just briefly looked at, this chapter will focus particularly on the fact that it does not include "non-capitalist modes of production." I believe that this characteristic of Marx's problem setting had a great influence on the direction in which others later argued about national disparities in labor wages. In other words, this has created the following two trends in this direction. One is to discuss "national differences of wages" while leaving "non-capitalist modes of production" out of the picture, as Marx did. From this approach, it can be said that a trend has emerged that places emphasis solely on the interpretation of the proposition "modification of the law of value" in Chapter 20 of Volume 1 of "Capital." [101] Another approach, just the opposite of what I have just described, is to take into account the issue of the "non-capitalist modes of production" but ignore the statement in Chapter 20 of Volume 1 of "Capital." I believe that both of these approaches are insufficient. Therefore, in this chapter, I would like to approach the subject while taking into consideration both the description in Chapter 20 of Volume 1 of "Capital" and the "non-capitalist modes of production." However, when we refer to "non-capitalist modes of production" in

---

101 Nawa (1949), although belonging to the first stream of commentators, focuses on the low cost of agricultural products produced by small farmers, which I will discuss at the beginning of Section 3 in this chapter. However, the influence of the existence of non-capitalist modes of production on wage levels goes beyond this point. And this is also what this chapter will show.

this chapter, we reaffirm the fact that the above-mentioned institutional condition of labor supply availability (freedom from servitude situation) is always attached to them.

Here is an overview of the structure of this chapter. Section 2 examines the proposition of "modification of the law of value" in Chapter 20 of Volume 1 of "Capital," which results in national differences in price levels. Section 3 examines Meillassoux's theory of labor supply from non-capitalist modes of production. Meillassoux's argument can be said to be a typical and standard discussion of the influence of labor supply from non-capitalist modes of production on wage levels. However, I think there are two missing points. One is that, as mentioned earlier, the description in Chapter 20 of Volume 1 of "Capital" is not considered. Another problem is that although Meillassoux's argument discusses a non-capitalist modes of production, it does not include a theory of community. Therefore, in Section 4 of this chapter, I will elaborate on the latter of these points. I will therefore consider what unique problems arise in relation to wage levels when non-capitalist modes of production are maintained by community. Section 5 then summarizes this chapter.

Before getting into the main text, I would like to make two points about the terminology used in this chapter.

First, in this chapter, in addition to developing countries, I also use the term latecomer countries as a similar concept, and distinguish between the two. Here, a latecomer country is a country that started developing its capitalist system later than a first-mover country, and therefore, unless there is a subsequent reversal, it represents a country with a low degree of capitalist development. In this case, latecomer countries are a concept related to their relative level of development, and although it may sound somewhat paradoxical, latecomer countries that are highly developed can also exist compared to countries that are even more developed. Characters of developing countries, on the other hand, are not only that the level of domestic capitalist development is low, but also that the majority of the population lives under non-capitalist modes of production. In other words, in the concept of a developing country, the framework of an economic structure consisting of multiple layers of production modes is within the field of view of the analyst.

Second, when we refer to wage simply below, we mean wage for simple labor. This is because the subject of this chapter is the elucidation of national differences in labor wages. This question naturally assumes that labor markets maintain uniformity within each country. Moreover, the simple labor market, rather than the labor market consisting of complex labor with special development, embodies the unity of the labor market because of its universality that exists across sectors.[102]

2. Price Levels and "National Differences of Wages"

Chapter 20 of Volume 1 of "Capital," "National Differences of Wages," points out that wage levels differ between countries with different degrees of capitalist development, as shown in the following quotation. It discusses high wages in developed countries with high price levels (that is, low relative value of money). This can be said to be the underlying reason for the wage disparity between first-mover and latecomer countries. Therefore, we will begin by considering this point.

"In proportion as capitalist production is developed in a country, in the same proportion do the national intensity and productivity of labor there rise above the international level. The different quantities of commodities of the same kind, produced in different countries in the same working-time, have,

102 I learned from Aramata (1972) that the "simple labor market" is a labor market that supports a "unified national labor market," although the nuance of the reason differs from mine.

therefore, unequal international values, which are expressed in different prices, i.e., in sums of money varying according to international values. The relative value of money will, therefore, be less in the nation with more developed capitalist mode of production than in the nation with less developed. It follows, then, that the nominal wages, the equivalent of labor-power expressed in money, will also be higher in the first nation than in the second; which does not at all prove that this holds also for the real wages, i.e., for the means of subsistence placed at the disposal of the laborer."

As can be seen, the conclusion that labor wages are higher in countries with a higher degree of capitalist development is clear, but the logic leading to this conclusion is likely to be twisted. This is because countries with a higher degree of capitalist development should have higher labor productivity. If this is the case, then if we make a simple deduction without taking into account the "modification of the law of value," which will be discussed later, the prices of the necessary means of subsistence should be lower there. In this case, if other conditions such as the workers' standard of living are held constant, wages would actually be lower. However, the quote above reaches a conclusion that is contrary to this "naive" reasoning. Therefore, it is necessary to explain the logic behind such a "reversal."

It may be possible to infer here that in countries with a high degree of capitalist development, workers have a high standard of living and therefore receive high wages. And to some extent, this can be said to be the case in modern times. However, as is clear from the quotation, this kind of reasoning is at odds with Marx at that time. This is because Marx disclaims that "which does not at all prove that this holds also for the real wages, i.e., for the means of subsistence placed at the disposal of the laborer." By doing so, he emphasizes that the high wages of workers in countries with a high degree of capitalist development described above are not a reflection of their high standard of living.

Immediately before this paragraph on national differences of wages is the following statement related to the so-called "modification of the law of value," which I have just touched upon.

"In every country there is a certain average intensity of labor, below which the labor for the production of a commodity requires more than the socially necessary time, and therefore does not reckon as labor of normal quality. Only a degree of intensity above the national average affects, in a given country, the measure of value by the mere duration of the working-time. This is not the case on the universal market, whose integral parts are the individual countries. The average intensity of labor changes from country to country; here it is greater, there less. These national averages form a scale, whose unit of measure is the average unit of universal labor. The more intense national labor, therefore, as compared with the less intense, produces in the same time more value, which expresses itself in more money.

But the law of value in its international application is yet more modified by this, that on the world-market the more productive national labor reckons also as the more intense, so long as the more productive nation is not compelled by competition to lower the selling price of its commodities to the level of their value."

It is clear from the positional relationship of the two descriptions that the "modification of the law of value" developed here brings about the "reversal" of "national differences of wages." But by what logic? This point will be considered below. Let us begin this work by understanding the content of "modification of the law of value."

However, in the order of things, let us first confirm the following points regarding the law of value concerning domestic labor intensity. Within a country, extra labor expenditure due to labor of less than standard intensity is not included in the calculation of commodity value. However, because the product of labor of standard intensity determines its value in the market, labor of above standard intensity appears as labor that forms value several times more than labor of standard intensity within the same period of time.

However, according to Marx, the law of value concerning labor intensity, which was valid domestically, needs to be "modified" in the following four points when applied to the world market.

① The average labor intensity of each country is summarized as the "national labor intensity," and the labor intensities in the world market are distributed as if the "national labor intensities" of each country form a stairs.

② There is no international standard labor intensity. In this case, a situation similar to the case of high-intensity labor within a single country arises, in which the ability of people with high labor intensity to create value in the same amount of time is high. However, not only that, the labor of people with low labor intensity can also be considered value-forming labor, although it is evaluated accordingly as labor with low value-forming power.

③ As mentioned in ②, although there is no standard labor intensity internationally, there is a unit of measurement that measures the "national labor intensity" of countries. It is the "average unit of universal labor."

④ All of these arguments, including those regarding intensity from ① to ③, as well as discussions regarding domestic conditions that are not subject to "modification," apply directly to "productivity." As a result, the labor of a nation with high productivity "reckons also as the more intense." However, it should be noted that there is a next reservation here; "so long as the more productive nation is not compelled by competition to lower the selling price of its commodities to the level of their value." This is because there is no such reservation regarding intensity.

Now, the content of the "modification of the law of value" is as described above, but there are actually some points there that are difficult to understand, and for this reason, the so-called "international value debate" has developed both domestically and internationally.[103] However, here I will not go into the individual points of the "international values debate" and examine this debate as such. This is because, while the "international value debate" was primarily concerned with trade theory, which is summarized in the proposition of comparative production costs, the subject of this chapter is a different issue; national differences of wages. Here, while learning from the arguments of various authors, I would like to limit my description to presenting my understanding of "modification of the law of value."

As a starting point for this work, let us consider the question of why a standard labor intensity exists domestically but not in the universal market. Domestic standard labor intensity is achieved through competition among workers, with high-intensity (intensive) labor eliminating low-intensity (relaxed) labor. (Muraoka 1976) Therefore, the fact that there is no standard labor intensity in the universal market means that competition among workers does not occur, at least not sufficiently. As a result, high-intensity (intensive) labor is not substituted for low-intensity (relaxed) labor. In other words, just because workers in country A are "lazy," they will not be replaced with "hard-working" workers in country B. This may occur because one of the following three conditions exists: (1) There are restrictions on the international movement of labor, (2) there are restrictions on the movement of capital, and (3) there are restrictions on both.

What I would like to pay attention to when making this three-way choice is the reservation mentioned above. In other words, although there is a possibility that national labor productivity may converge internationally, there is no

103 Regarding the international value debate, see Kinoshita (1960) and Naruse (1985).

mention of such convergence regarding national labor intensity. It may be said that the possibility of international convergence of national labor intensity is denied there, albeit implicitly. Or, if we are more cautious and do not reason this far, we can say the following. When comparing the national labor intensity and the national labor productivity, Marx believed that the latter would be easier to converge internationally. The following circumstances seem to be behind this difference in international convergence between national labor intensity and national labor productivity. The national labor intensity is largely determined by the qualifications of workers. On the other hand, while some of the factors that determine the national labor productivity depend on the qualifications of workers, such as their skill, they also largely depend on the performance and characteristics of objective production factors such as machinery and equipment.[104] Therefore, if Marx believed that the national labor productivity tends to converge internationally, but that such tendency is relatively weak in national labor intensity (I assume that Marx thought this way), then from here the following proposition is derived. Marx believed that while there were relatively strong political restrictions on the international movement of labor due to national borders, there were relatively weak restrictions on the international movement of objective means of production.[105]

By the way, among the objective factors of production, land naturally

104 "This productiveness is determined by various circumstances, amongst others, by the average amount of skill of the workmen, the state of science, and the degree of its practical application, the social organization of production, the extent and capabilities of the means of production, and by physical conditions." (Marx 1867)

105 In this way, Muraoka (1976, 1988) sees the structural characteristics of the world market in the political restrictions on the movement of labor through national borders. However, he sees the determinant of national labor productivity disparities solely in objective conditions of production. On the other hand, I also emphasize skill level as a determining factor for productivity disparities.

cannot be moved internationally, but differences in the fertility of land between countries also determine differences in the productivity of national labor because of their great influence on agricultural productivity. Earlier, Marx gave the connotative expression, "so long as the more productive nation is not compelled by competition to lower the selling price of its commodities to the level of their value," but he did not conclude that "because the more productive nation will lower the selling price," without making a definitive statement. This seems to be due to consideration given to the fact that in addition to labor, there are objective production factors such as land that are even more difficult to move internationally than labor. On the other hand, for other objective production factors such as machinery and equipment, it can be assumed that a country's domestic capital imports them from another country and brings them into the country, or that foreign capital takes the lead in introducing them.

For this reason, the labor of countries with high national labor intensity and national labor productivity produces more value in the world market in the same amount of time than the labor of countries with low national labor intensity and productivity. In other words, it is multiplied labor. In terms of the value produced, for example, one working-day in country A is equivalent to three working-days in country B.[106] In this case, the product of country A in one working-day has the same value as the product of country B in three working-days. It goes without saying that country A, which has a high national labor intensity and productivity, is a first-mover country, and country B, which has a low national labor intensity and productivity, is a latecomer.[107]

By the way, so far I have been casually using the concept of the national

106 This numerical example is from Marx (1862-1863)

107 It can be said that the national labor intensity is higher in advanced countries due to the introduction of modern labor management.

labor productivity, which is said to have originated from Marx, but as many commentators have pointed out, this is actually a concept that is difficult to grasp. On the other hand, as mentioned earlier, the national labor intensity tends to converge to a standard one through competition among workers, and in that sense it has a reality. However, since productivity is a concept related to the amount of use-value produced per unit of labor time, it is impossible to quantitatively compare productivity between production departments whose products have different use-values. For example, disparities in the productivity of agricultural labor for the same crop can be questioned, but even disparities in the productivity of different agricultural crops cannot be strictly questioned. It goes without saying that the disparity in productivity between agriculture and industry cannot be a problem. Since it is difficult to even compare productivity across different sectors, it must be even more difficult to aggregate the productivity of each sector into the productivity of national labor (for example, while applying some kind of mathematical processing such as averaging the productivities of different departments within a country). Despite this, Marx is said to have suggested the existence of the concept of the national labor productivity.[108] What would that be ? In fact, how to solve this difficult problem seems to have been the central issue in the "international value debate." Here, as mentioned earlier, I will not get too hung up on this controversy, but will simply present my own "solution" method.

In order to compare and aggregate productivity across different departments, it is necessary to establish a common scale for measuring productivity across departments. However, since there is generally no commonality between sectors in the use-value of goods, such a common measure cannot be related to use-value. By the way, it goes without saying

108 However, as can be seen in the quotation, this concept is not used in its literal form.

that the two elements of a product are use-value and value, so a common measure that has nothing to do with use-value must be one that measures the magnitude of value. Therefore, by using efficiency as an alternative indicator of productivity, it is possible to approximately transform productivity, which is originally a use-value concept, into a value concept. In this way, it becomes possible to obtain a common scale for approximately measuring productivity disparities between sectors. This makes it possible to calculate national labor productivity as the average labor efficiency of various sectors within a country.[109] Furthermore, rather than using a simple average of the labor efficiency of various sectors, it would be more appropriate to use the following weighted average to reflect the weight of each sector in the national economy.

$$Pn = \Sigma\ (Pi \times Li\ )\ /\ \Sigma Li$$

Where, Pn: national labor efficiency, Pi : average labor efficiency in sector i, Li : number of workers in sector i.

As is clear from the reasoning above, national labor efficiency is an alternative indicator calculated to approximately aggregate productivity at the national level.[110] However, if we examine the differences between countries in the efficiency of national labor in detail, we find that they do not merely represent the differences between countries in the productivity of national

109 As mentioned in the previous note, Marx does not use the word productivity (*Produktivkraft*) in the quotation, but he does use the word efficiency (*Produktivität*).

110 Muraoka (1976) considers the national labor productivity gap to be the weighted average of the sectoral productivity gap. However, for the reasons discussed earlier in this chapter, it may be difficult to calculate this. Nawa (1949) uses the productivity of key industry to represent the productivity of national labor. However, this makes it impossible to determine the level of national labor productivity for countries that do not have a key industry. Today, it would be the automobile industry.

labor. In other words, national differences in national labor efficiency also include national differences in national labor intensity.

Let me explain that. Domestic standard labor intensity permeates across sectors, so differences in efficiency between sectors do not reflect differences in intensity between sectors. More precisely, under an ideal situation in which labor intensity converges to a standard at the national level, there would be no difference in intensity between sectors. Therefore, the domestic efficiency gap between sectors only approximately reflects the "productivity gap between sectors." However, differences between countries obtained by comparing national labor efficiency, which is an aggregate value at the national level, do not only approximately reflect differences in national labor productivity. Since labor intensity does not converge internationally, they also reflect national differences in labor intensity.

Now, using the concept of national labor efficiency, we can rephrase what was stated above as follows. The labor of countries with high national labor intensity and national labor productivity appears on the world market as intensified labor that produces more of the same type of product and value in the same amount of time than labor of countries with low national labor intensity and productivity. This can be approximately expressed as national differences in national labor efficiency. For example, one working-day in country A is equivalent to three working-days in country B.

By the way, I just casually mentioned "equivalent," but if I were to express this more concretely based on the current example, it would be as follows. The value of country A's product in one working-day is equal to the value of country B's product in three working-days. And in order for this to be not just a fantasy but something that can actually be confirmed by actual measurement of value, there must be some means of measuring the value of these products. Needless to say, something that functions as such a measuring instrument would be money, which is the universal equivalent. And in order for it to function universally as the measure of values for the products of labor, it must have a constant value for the same quantity everywhere, in our case in country A and country B. Thus, the premise of the relationship that “one working-day in country A is equivalent to three working-days in country B” is the existence of the universal equivalent (money) that has a constant value everywhere for the same quantity.

The reason why I go out of my way to emphasize this is because I want to emphasize the difference in the mechanism of value formation between money-commodity (hereinafter referred to as gold) [111] and general products. For general commodities, the market-values (hereinafter simply referred to as domestic market-values) are first determined in the domestic market of each country.[112] The international market-value is then determined as the international weighted average of these values. Therefore, the domestic market-value of each country can be a different amount from the international market-value. This discrepancy in market-values may be due in part to the existence of barriers defined by political restrictions on labor movement between the domestic market and the international market. As a result, competition in universal market is incomplete for many commodities. Nawa (1949) discusses this as follows.

“All domestic commodities belonging to various industrial sectors do not directly participate in the universal market, where their values are

111 The following is Matsumoto’s (1995) argument regarding the functions of gold as measure of value. “Even under the suspension of convertibility, if we focus on the market price of gold, which fluctuates daily, it can be argued that gold functions today as the standard of price, and therefore as the measure of value.”

112 “On the one hand, market-value is to be viewed as the average value of commodities produced in a single sphere, and, on the other, as the individual value of the commodities produced under average conditions of their respective sphere and forming the bulk of the products of that sphere.” (Marx 1894)

immediately determined according to universal market-values and international values. The opposing relationship between individual value and social value for all products does not appear as a direct opposition between individual value and universal market social value. The individual values of many products are summarized as national and social value, and after being determined as value and price, it appears on the universal market. Of course, it is also possible that there are commodities that are separated from the universal market."

Furthermore, let us consider that the individual values of general commodities that are imported and exported immediately participate in the formation of international market-value without going through the intermediary of domestic market-value formation as described above. In this case, each country's domestic market-value system would be denied. This is inconvenient. (Matsumoto 1995) If the international market-value were to be accepted as the domestic market-value, it would be different from the domestic amount of labor invested, so the law of value, which sees equal labor invested behind equal value, cannot be applied domestically. This contradicts the above-mentioned premise regarding restrictions on international movement of labor. This is because, under this premise, the law of value should be pervasive within a country. For these reasons, Prof. Nawa's view of the universal market, which states that "each nation is an integral part," seems appropriate.

"Each country is assumed to be a closed economy with its own production system and its own social division of labor. Starting from this, countries trade with each other internationally and can only have economies in society, not in isolation. This is different in nature from the complete division of labor between the producers of each commodity." (Nawa 1949)

Let us consider in more detail the situation where domestic market-value and international market-value diverge. For example, assume that the domestic market-value of one unit of commodity (I) is equivalent to one working-day in country A, and six working-days in country B. Furthermore, we assume the above-mentioned relationship: 1 working-day in country A = 3 working-days in country B. In this case, if the international market-value is determined as the weighted average of the domestic market-values of countries A and B, then the domestic market-value of country B will be included in it as twice (not 6 times) the domestic market-value of country A. As a result, if both countries produce the same amount of the commodity, the international market-value, which is the weighted average of the domestic market-values of both countries, will be as follows. That would be the equivalent of 1.5 working-days for country A and 4.5 working-days for country B (**Table 5-1**). [113]

Unlike this general commodity, gold, which is the money-commodity, is the unique commodity whose international market-value is determined by a small number of gold-producing countries. For this reason alone, for many countries other than gold-producing countries, the discrepancy between domestic and international-market values is not a problem, as is the case with general commodities. Gold mines around the world compete directly with each other because gold has a high value relative to its weight and transportation costs are low. (Smith 1776) Regarding the characteristics of the circulation of money-commodity, I would like to introduce the following passage from Volume 1 of "Capital." There are two flows; (1) from countries of origin to universal market, and (2) movement between countries mediated by foreign exchange transactions and exchange rate fluctuations.

"The current of the stream of gold and silver is a double one. On the one hand, it spreads itself from its sources over all the markets of the world, in order to become absorbed, to various extents, into the different national spheres of circulation, to fill the conduits of currency, to replace abraded gold and silver coins, to supply the material of articles of luxury, and to petrify into

---

113 (1+2) /2=1.5. (3+6) /2=4.5.

Table 5-1 Discrepancy between domestic market-value and international

		Country A (First-mover country)
Commodity (Ⅰ) (Advanced industry)	Domestic market-value	1 working-day/unit (Gold 1g)
		∧
	International market-value	1.5 working-days/unit (Gold 1.5g)
Commodity (Ⅱ) (Non-advanced industry)	Domestic market-value	1 working-day/unit (Gold 1g)
		∨
	International market-value	0.75 working-days/unit (Gold 0.75g)

Note 1) It is assumed that the production volumes of commodities (Ⅰ) and (Ⅱ) are equal in both countries A and B.

hoards. This first current is started by the countries that exchange their labor, realized in commodities, for the labor embodied in the precious metals by gold and silver-producing countries. On the other hand, there is a continual flowing backwards and forwards of gold and silver between the different national spheres of circulation, a current whose motion depends on the ceaseless fluctuations in the course of exchange."

However, as Nawa (1949) points out, Marx in this passage does not address the issue of differences in fertility in precious metal mines. On the other hand, in the chapter discussing rent in mining in Volume 3 of "Capital," Marx states that "mining rent proper is determined in the same way as agricultural rent." He suggests that differences in the fertility of mines participating in metal production should be taken into account in forming metal prices, and that prices are determined by inferior mines at the production limit.

However, Smith wrote in "The Wealth of Nations" (1776) that "the price of every metal at every mine, therefore, being regulated in some measure by its price at the most fertile mine in the world that is actually wrought." There, he may be advocating for metal prices to be determined by high-quality mines.

market-value

Country B (Latecomer coutry)	
6 working-days/unit (Gold 2g) ∨ 4.5 working-days/unit (Gold 1.5g)	The apparent productivity gap between countries A and B (6 times) exceeds the national labor productivity gap between the two countries (3 times) .
1.5 working days/unit (Gold 0.5g) ∧ 2.25 working-days/unit (Gold 0.75g)	The apparent productivity gap between countries A and B (1.5 times) is lower than the national labor productivity gap between the two countries (3 times) .

2) 1 working-day in country A = 3 working-days in country B = 1g of gold

However, Smith's next sentences state that the value of precious metals is determined by the quantity theory of money, based on the quantity of precious metals supplied, the quantity of general products that money should make circulate, and the relationship between these two. The meaning of the previous quotation must also be understood with this in mind.

"The quantity of the precious metals may increase in any country from two different causes; either, first, from the increased abundance of the mines which supply it; or, secondly, from the increased wealth of the people, from the increased produce of their annual labor. The first of these causes is no doubt necessarily connected with the diminution of the value of the precious metals, but the second is not."

It is thought that these arguments of Smith were strongly connected to his understanding of the situation regarding the discovery of precious metal mines in America in the 16th century and the subsequent worldwide decline in the value of precious metals. In other words, Smith believed that the discovery and exploitation of these rich mines increased the supply of precious metals relative to the amount of general produce circulating in the countries, leading to a decline in their value from the 16th century onwards. The aforementioned proposition of metal value regulation based on superior mines is considered to be a comprehensive expression of this situation. Considering the following, Smith's argument can coexist without contradiction with Marx's earlier proposition that "mining rent proper is determined in the same way as agricultural rent." That is, the discovery and exploitation of rich mines will result in the displacement of marginal precious metal mines, provided that the global demand for precious metals does not increase as much as the increase in its supply. This results in a decline in the value of precious metals. Therefore, if we adopt the rule of gold value based on rent theory, the international market value of gold is not determined by the weighted average of the labor expenditure per weight of gold of each mine participating in gold production in the gold-producing countries. Rather, it will be regulated by the individual value of gold produced in mines at the production threshold. Here, the "small number of gold-producing countries" that determine the international market value of gold are specifically defined as countries that have gold mines at the production limit. Such a mine would normally be a single mine, as is clear if we examine the meaning of the term production limit. If multiple mines with the same fertility reach the production limit, then there will be multiple mines, but this is not a general case. Thus, the "small number of gold-producing countries" is usually a single country, although the term "countries" is used because there may be multiple countries. However, what we must further consider here is the question of how the production limit point is specifically determined in the world market. According to Sato (1983-1984), this is not a production limit due to simple differences in fertility such as inside one country. First, it is the lowest quality mine after undergoing the "modification of the law of value." Furthermore, it is the lowest quality mine, taking into account differences in absolute land rent levels between countries. However, since this chapter does not discuss the international market-value of gold as such, I will simply point out these points and proceed with the discussion in line with the main theme of this chapter.[114] It suffices here to observe that the international market-value of gold is regulated by the individual value of gold produced at the global production limit, usually in a single mine.

By the way, in the previous example, the discrepancy between the domestic market-value and the international market-value is such that in country A, the domestic market-value is less than the international market-value, and in country B, the domestic market-value exceeds the international market-value. However, it is clear that this magnitude relationship between domestic market-value and international market-value occurs under specific conditions of productivity as described below. This magnitude relationship is the result of the apparent difference in productivity between countries A and B regarding the commodity in question, which does not take into account the "modification of the law of value," and which exceeds the difference in the productivity of national labor between the two countries. Let us try to imagine a completely opposite situation. For example, if the domestic market-value of one unit of another commodity (Ⅱ) is equivalent to one working-day in country A and 1.5 working-days in country B, then the apparent productivity gap for that commodity is 1.5 times. This is smaller than the three times the national labor productivity gap previously assumed between the two countries. In this case, when the international market-value is determined, the domestic market-value of country B will be 0.5 times the domestic market-value of country A, not 1.5 times. The average international market-value of the commodity in countries A and B is equivalent to 0.75 working-days in country A and 2.25

114 For a more detailed explanation of these points, see Sato (1983-1984). Sato (1994) also provides an interesting discussion of the Heckscher-Ohlin theory, which emphasizes differences in commodity demand and factor endowments between countries in the formation of the international division of labor.

working-days in country B, assuming equal production in both countries(**Table 5-1**). Therefore, in this case, the domestic market-value of country A's commodity (II) exceeds the international market-value, contrary to the aforementioned commodity (I), and the domestic market-value of country B is less than the international market-value. In summary, while the national labor productivity gap is used as the standard, the relationship between domestic market-value and international market-value differs depending on whether the apparent productivity gap in a certain sector is larger or smaller. In other words, if between two countries, the apparent productivity gap in a certain sector (I) exceeds the national labor productivity gap, the domestic market-value of the country with high national labor productivity (first-mover country) is below the international market-value. Conversely, in country where the productivity of national labor is low (latecomer country), the domestic market-value exceeds the international market-value. On the other hand, if the apparent productivity gap in a certain sector (II) between two countries is less than the national labor productivity gap, the domestic market-value of the first-mover country will exceed the international market-value. Conversely, in latecomer country, domestic market-value is lower than international market-value. Furthermore, since it can be said that the more advanced the industrial sector is, the larger the difference in productivity between first-mover country and latecomer country is, it would be appropriate to define (I) as an advanced industry and (II) as a non-advanced industry.

By the way, as mentioned earlier, the international market-value of money commodity (gold) does not deviate from the domestic market-value for normal national economies, except for "a few gold-producing countries." There international market-value is a single given.[115] Since it is a given value, the

115 In gold-producing countries that do not determine the international market-value of gold, the discrepancy between the domestic market-value of gold and

value of gold always reflects the disparity in the productivity of national labor. Therefore, let us now assume that there is a relationship where the value of 1 gram of gold = 1 working-day in country A = 3 working-days in country B. Incidentally, the gold value in this case, as mentioned earlier, is the individual value of gold produced in mines that are at the global production limit. Thereby, the weight of gold, or price, that represents the value of one unit of commodity (I) (advanced industrial product) is 1g in terms of the domestic market-value (price) of country A. Also, the domestic market value (price) of country B is expressed in 2g of gold, and the international market value (price) is expressed in 1.5g of gold. On the other hand, the price, which is the weight of gold that represents the value of commodity (Ⅱ) (non-advanced industrial product), is 1g in country A, 0.5g in country B, and 0.75g in the international market. In short, for commodities from non-advanced industries, the value (price) of first-mover country A is rather expensive compared to the international market-value (price), whereas for commodities from advanced industries, this relationship is reversed. In other words, the value (price) of first-mover country A is rather cheap compared to the international market-value (price). The fact that Marx included agriculture among these non-advanced industries is clear from the following quotation in "History of the Theory of Surplus-Value" (1862-1863).

"In the case of manufactured goods, it is well known that a million people in England produce far more products than, say, those in Russia, even if the individual goods are much cheaper, and produce products of far greater value. However, in the case of agriculture, the same relationship does not seem to exist between capitalist-developed nations and relatively undeveloped nations. The products of people who are lagging behind are cheaper than the products

---

the international market-value may become a problem. The following discussion covers non-gold-producing countries.

of people who have developed under capitalism. This is true from the point of view of *monetary prices*. Nevertheless, the products of developed nations appear to be the products of much less labor (in a year) than those of lagging nations."

The "civilizing effect of capital" naturally extends to agriculture, and the agricultural productivity of first-mover country becomes higher than that of latecomer country. In first-mover country, the differentiation and disintegration of the peasant classes progresses, and agricultural management bodies that embody advanced agricultural productivity are formed. However, on the one hand, due to the technical conditions specific to agricultural production, and also because the surplus-profits generated in agriculture are absorbed into land ownership in the form of rent, which constrains the investment of capital in this sector, the development of agricultural productivity lags behind industrial productivity.[116] As a result, the gap in agricultural productivity between first-mover and latecomer countries is not as wide as the gap in industrial productivity between the two. This is probably what Marx is trying to say there.[117]

---

116 "If the composition of capital in agriculture proper is lower than that of the average social capital, then, *prima facie*, this expresses the fact that in countries with developed production agriculture has not progressed to the same extent as the processing industries. Such a fact could be explained — aside from all other circumstances, including in part decisive economic ones — by the earlier and more rapid development of the mechanical sciences, and in particular their application compared with the later and in part quite recent development of chemistry, geology and physiology, and again, in particular, their application to agriculture." (Marx 1894)

117 Nowadays, even though they are developed countries, the New World countries of North America and Australia have developed large-scale farming based on vast farmland and are achieving high agricultural productivity. On the other hand, some developing countries are progressing in industrialization through foreign direct investment by multinational companies. These current circumstances will force us to revise, or at least partially reconsider, the relationship that Marx pointed out between the disparity in agricultural and

The above is consistent with the comparative production cost proposition derived from Ricardo (1817), which states that the domestic market-value (price) of commodities are lower in comparatively advantageous sectors with relatively high productivity in each country. In first-mover country, the absolute level of productivity is high regardless of whether it is an advanced or non-advanced industry. However, this will be reflected in the level of domestic market-value (price) after going through a comparative production cost structure. Therefore, even in latecomer country, in sectors where they have a comparative advantage, domestic market-values (prices) are lower than those in first-mover country.

By the way, a country's price level (relative value of money) is a kind of average of the domestic market-values (prices) of various commodities. However, in the current context where we are concerned with national differences in labor wages, we should not be concerned with commodities in general. The commodity types that should be noted here are those centered on consumer goods for workers. It can be said that the types of commodities that make up consumer goods for workers include many products from non-advanced industries such as agricultural products and light industrial products. By the way, as we have just seen, domestic market-values (prices) vary in magnitude between first-mover country and latecomer country, depending on whether they are due to advanced or non-advanced industries. Domestic market-values (prices) of commodities from non-advanced industries are lower in latecomer country. It thus follows that in first-mover country the price level (the reciprocal of the relative value of money) for workers is high, and therefore wage is high in first-mover country as long as wage reflects the prices of the necessary means of subsistence. Conversely, wage is low in

---

industrial productivity between first-mover and latecomer countries. This point is emphasized by Motoyama (1982).

latecomer country for the opposite reason. The effects of living standards and their height on wage, which were not discussed here, will be discussed later in Section 5.

By the way, the discussion in this section has been conducted at the level of value theory. On the other hand, Bauer (1907) and Grossmann (1929) have discussed the application of theory of price of production on a global scale.[118] Bauer and Grossmann saw the exploitation of the latecomer country by the first-mover country in the formation of global average profit and the transfer of surplus-value from the latecomer country to the first-mover country through the distribution process. Nawa (1949) offers a methodological criticism of the Bauer-Grossmann theory. The point he points out is that the following three prerequisites for applying the theory of price of production are not satisfied in the world market. First, capitalist relations are completely dominant in all industrial sectors in each country. Second, there is freedom of movement of workers internationally. Third, there is free movement of capital internationally. Although it may be necessary to reserve this criticism regarding the third condition today, the other two points seem to me to be valid. Therefore, I agree with his opinion that it is impossible to apply the theory of price of production to the world market. In this way, it is confirmed that the discussion in this section, which deals with the following two issues, needs to be conducted at the level of value theory, rather than the theory of price of production. The two problems are the difference in the value-forming power of national labor between first-mover country and latecomer country, and the resulting national differences in price levels and labor wages.

However, the question is how the absence of the first condition in the world market will affect the "national differences in labor wages" that are the subject

118 Emmanuel (1969) inherited their arguments after World War II.

of this chapter. It would be insufficient to simply consider this question as a condition for applying the theory of price of production. Perhaps this requires more treatment than that. In other words, as stated at the beginning of this chapter, it is necessary to further examine the question of what kind of consequences the non-capitalist modes of production that exists in developing countries will have on wage levels there. Therefore, consideration of this point will form the central point of discussion in the following sections.

3. Low Wages in Developing Countries through the Maintenance and Dismantling of Non-Capitalist Modes of Production

1) Inexpensiveness of Agricultural Products Produced by Peasants

We will consider the question of what kind of influences the non-capitalist production modes, which currently exist in developing countries, have on wage levels there. First, let me present the following passage, which is often quoted from "Capital," Volume 3, Chapter 47, "Genesis of Capitalist Ground-Rent."

"For the peasant owning a parcel, the limit of exploitation is not set by the average profit of capital, in so far as he is a small capitalist; nor, on the other hand, by the necessity of rent, in so far as he is a landowner. The absolute limit for him as a small capitalist is no more than the wages he pays to himself, after deducting his actual costs. So long as the price of the product covers these wages, he will cultivate his land, and often at wages down to a physical minimum." "This is one of the reasons why grains prices are lower in countries with predominant small peasant land proprietorship than in countries with a capitalist mode of production. One portion of the surplus-labor of the peasants, who work under the least favorable conditions, is bestowed gratis upon society and does not at all enter into the regulation of price of production or into the creation of value in general. This lower price is consequently a result of the producers' poverty and by no means of their labor productivity."

In a free competitive society where the capitalist mode of production is dominant, and where most agricultural production is carried out under capitalist management, the price of agricultural products in the market is determined by the cost of the marginal product that satisfies social demand. Furthermore, agricultural product price is formed by adding average profit and absolute rent to this cost. Differential rents are added to this amount for agricultural products produced under more favorable conditions. However, in places where the degree of capitalist development of society is low and agriculture is generally carried out by peasants, these economic categories have not yet been socially established. Peasants can continue agricultural production if they can secure enough home labor wages through the sale of agricultural products to barely maintain their livelihood. The prices of agricultural products thus supplied can be lower than those of capitalist agriculture, but this is not due to the high productivity of agricultural labor, but due to the peasants' distressed sales.[119] Furthermore, when agricultural product prices are determined by policy, some kind of bias will be given to the prices determined in the markets based on certain policy goals. However, even in that case, it is impossible to arbitrarily set prices that exceed market prices, and market prices have a meaning as a standard for policy prices. (Kurihara 1955)

Naturally, the lower prices of agricultural products resulting from peasants' production lead to a decline in the value of labor-power, which in turn becomes a factor in lowering the wages that workers receive. However, the fact that

---

119 Hazama (1953) argues that the low price of small-scale peasants' products cannot be explained solely by their tolerance for suffering even if they do not receive all of their income. More positively, he argues that this stems from the fact that peasant labor cannot assert itself as social labor because it has the character of lesser labor of lower intensity.

capitalist agricultural management is not common is not a special situation found only in developing countries, but is a common phenomenon even in modern developed capitalist countries. Therefore, such distressed sales by peasants can be seen even in developed countries, and to that extent, it cannot be said to be a special factor that brings about low prices for agricultural products in developing countries.

However, if we go further and consider the extent of distress sales, the situation differs between developed and developing countries in this respect. That is, in developed countries, the degree of categorical establishment of C (constant capital) and V (variable capital), which constitute cost-prices, is higher. Therefore, although it is possible that V changes elastically to some extent, agricultural product prices are formed based on such cost-prices of marginal products. In contrast, in developing countries, the degree of categorical establishment of C and V is relatively low. Therefore, as Marx said, agricultural prices are sometimes even set based on the minimum physical labor wage of peasants who produce marginal products (let this wage be v now). If peasants do not specialize in agriculture and instead work in various occupations, V or v included in agricultural product prices can be further reduced. This is because the income obtained from such side jobs can replace a portion of V or v included in agricultural product prices. Part-time work can be seen even in the low stages of development of a commodity economy, where the economy is semi-subsistence-oriented, and conversely, it can also occur in the process of differentiation of peasants as a result of the penetration of the commodity economy into rural areas. In any case, however, this situation probably characterizes rural villages in developing countries rather than in developed countries, where occupational differentiation is advanced and agriculture tends to become a specialized occupation.

Furthermore, when the prices of marginal agricultural products are set at the level resulting from the previous distress sales and these come to regulate the market prices of agricultural products, agricultural products produced under more advantageous conditions will generate differential rents.[120] If the peasant is a landed farmer, the differential rent will become his own income. (Marx 1894) In any case, the prices of agricultural products in developing countries can be lowered not only by the comparative advantage of their agricultural productivity as seen in the previous section, but also by such distressed sales by peasants.

However, as shown in the earlier quotation from Marx, these arguments apply when the peasant is a free owner of his own land. In reality, land ownership exists as a landowner class in conflict with peasants, and there are situations in both developed and developing countries where peasants rent land from the landowners and carry on farming. In that case, it is not "society" but "landowners" who receive the surplus-labor of peasants in marginal production conditions. Agricultural product prices must include the rent paid to the landowner class, and will therefore rise accordingly. Since this rising price is a result of land ownership preventing peasants from farming freely, if it becomes fixed, it becomes a kind of rent with characteristics similar to

---

120 In places where peasants' production is dominant, land rent is only differential rent, and there is no room for absolute rent. (Marx 1894) Just to be clear, peasant's differential rent is the surplus formed by peasants with good production conditions, and is not capitalist rent, which is based on the establishment of the price of production. In the case of peasants, the factors that determine the quality of production conditions include not only the fertility and location of the land, but also the area under management, production equipment, and the quantity and quality of family labor. (Tashiro 1968) Therefore, in this chapter, peasants who determine market prices are not expressed as existing on marginal land, but as existing under marginal production conditions. Therefore, in peasants' production, even if the farm is located on good land, it may be a marginal farm that determines market prices. I learned about the relationship between agricultural product prices and land rents from a series of papers published in Inuzuka (1982).

absolute rent.[121]

In this way, peasants in developing countries can continue agricultural production if they receive the minimum physical wage (v). However, this is not a factor in lowering agricultural product prices when there are constraints due to land ownership. In this case, the price of agricultural products must reach a level where the rent, which has the character of absolute rent, which falls into the pocket of the landowner, is added to the v of the peasant who produces the marginal product. Otherwise, peasants' participation in production will not take place. In this case, the landowner who owns land with more advantageous production conditions will receive not only rent like absolute rent but also rent like differential rent. If we consider the social existence of landowners in this way, the fact that peasants in developing countries can accept wages (v) at the lowest physical limit does not become a factor in lowering agricultural prices. In fact, it is a factor that could cause land rents to soar. In this case, the extreme hardship faced by peasant in developing countries cannot necessarily be said to be a factor in lowering agricultural prices and labor wages there. On the other hand, when the landowner class is wiped out through agrarian reform and peasant land ownership becomes socially established, agricultural prices decline before and after the reform. One example is none other than Japan, which underwent the reform after World War II. (Ishiwata 1959)

However, the influence of the non-capitalist modes of production on labor wage level is not limited to the reduction in agricultural prices that we have just seen. As mentioned earlier, non-capitalist modes of production can influence the wage level by partially bearing the reproduction costs of labor.

121 This kind of rent, which has characteristics similar to absolute rent, does not presuppose the social establishment of prices of production, but is surplus that arises from the soaring prices of agricultural products as a result of land ownership preventing peasants from farming freely.

Below, we will consider this point by examining the arguments of Meillassoux (1975), which are often cited when this type of problem is dealt with.

2) Non-Capitalist Modes of Production that Partially Bear the Costs of Reproducing Labor-power

Meillassoux's "*femmes, greniers et capitaux*" consists of two parts that are highly independent from each other. The first part is an economic anthropological study of families in subsistence economies, focusing on the functions of material production and human production that these families have. However, the second part takes a turn and deals with the issue of sources of low-wage labor originating from rural areas for the capitalist sector in modern developing countries. Of these, the second part, which is directly related to the subject of this chapter, will be the subject of discussion after introducing its contents in this section. The first thing to point out is that Meillassoux arguments in his book were based on his experience of research in various parts of West Africa, which he had been conducting since the late 1950s. Therefore, the geographical and temporal scope of the discussion on labor sources is limited to West Africa from the mid to late 20th century, or even sub-Saharan Africa if expanded a little. Alternatively, Meillassoux's arguments should at least partially reflect the special circumstances of these regions at that time. In fact, Meillassoux direct problem is to provide an explanation for low wages in colonial West Africa. This is because there was a mysterious situation that required clarification. "Despite the chronic labor shortage in West Africa, the remuneration of labor did not reflect this situation," meaning that wages were lower than in developed countries.

In addressing this problem, Meillassoux focuses on the concept of "labor reproduction costs," which consists of the following components: ① Living maintenance costs during the working periods of workers (immediate reproduction costs of labor-power), ② living maintenance expenses during non-working periods (unemployment, illness, etc.), ③ renewal costs (or reproduction costs) of the workers due to raising the workers' children.

In capitalist societies, direct wage payments to workers are usually only payments for this first element. Therefore, in order for the value of labor-power, which has the three elements mentioned above, to be fully paid and for labor-power reproduction to take place completely, direct wage payments alone are insufficient. Therefore, there is a need for indirect wage payments through the social security system to supplement direct wage payments. On the other hand, when workers are paid only direct wages, as was the case in Europe in the past and in many developing countries today, the maintenance and reproduction of labor-power cannot be guaranteed within the realm of capitalist production. It is inevitably projected into other modes of production.

By the way, there are two forms of labor movement from the non-capitalist sector to the capitalist sector. The first is the form of outflow from rural areas (*exode rural*), and the second is the form of return migration (*migration tournante*). According to Meillassoux, the first form was what Marx envisioned under the name of primitive accumulation, and was also the driving force behind the expansion of all developed capitalist economies, including Japan. The second form was common in post-World War II Africa and, moreover, "was the source of the huge and ever-increasing population movement between Africa and Europe." In this form, the self-sufficient family economy (*éconimie domestique*) is preserved, but the labor force is temporarily transferred to the capitalist sector and then returned to its source. Through this, the capitalist sector comes to "exploit" the non-capitalist subsistence family economy. The capitalist sector uses the labor-power produced in the family economy without paying for its support, and throws it back into the family economy when it is not in use. Through this operation, the capitalist sector is able to impose on the family economy the costs of reproducing workers over generations and the costs of maintaining their living during their non-working period, which are the components of the labor reproduction cost. Also, at least some of the worker's immediate reproductive costs can be spared if he "remains near the granary with his wife who prepares his daily meals." Meillassoux calls this value transferred from the family economy to the capitalist sector "rent in labor" (*rente en travail*). Furthermore, Mies et al. (1995) emphasizes the significance of unpaid work by "housewifed" women in this case, and applies the concept of "continuing original accumulation" to the resulting transfer of value to the capitalist sector.

However, in order for the capitalist sector to sustainably enjoy the "rent in labor," the familial reproduction of labor-power must not be endangered by a partial outflow of people to the capitalist sector. Therefore, it is necessary to organize the outflow of essential labor in such a way that it is limited to the agricultural off-season and returns to rural areas during the agricultural busy season. Furthermore, in order to preserve subsistence agriculture and the family economy, "paradoxically, capitalists must prevent the expansion of capitalism into the rural areas that provide the labor." For example, in the "reserves" developed in British settler colonies, land was systematically excluded from private ownership. On the other hand, in remote colonized areas unrelated to the development of export crops, "natural" reserves have been formed spontaneously that are not threatened by private land ownership. The African population living under subsistence agriculture in these remote areas has nothing to sell other than their own labor, and "the need for money (for paying taxes, purchasing local goods that were once bartered, substituting manufactured goods for handicraft goods, etc.)" forces them to enter the labor market.

By the way, as the discussion progresses to this point, the image that

Meillassoux has of a self-sufficient family economy becomes quite clear. In this case, the self-sufficient family economy is not completely unrelated to the money economy, but is becoming partially involved in the money economy due to the need to make payments and purchase manufactured goods. However, it basically has nothing to sell other than the labor of its family members, and as a result, it is forced to interact with the labor market. However, since there are corresponding sellers for "purchasing local goods," most of the products of the "self-sufficient family economy" are for home consumption, but some are sold. Therefore, a commodity economy based on these materials is also being operated on the side.

Let us return to introducing Meillassoux's argument.

In order to extract "rent in labor" in Europe, it was necessary to create unique institution, ideology, and mechanism. These are the dual labor market, racist ideology, and the rotation of immigrant workers. Immigrant workers from developing countries are placed in low-wage, unstable employment in Europe, and are also paid discriminatory indirect wages. As a result, they are included in a different category of labor market than workers who are fully integrated into the capitalist sector. Discriminatory ideology such as "racism and xenophobia" makes this dual labor market function. And in the case of illegal immigrants, their smuggling is "treated with expectation and leniency," since this in itself gives employers reason to devalue the migrant workers' working conditions. In this way, migrant workers who are deprived of social security and job security are ultimately forced to return to their home countries.

However, the self-sufficient family economy gradually declines due to the contradictions inherent in excessive exploitation by the capitalist sector, which we will discuss next. Ultimately, it loses the ability to continuously supply labor to the labor market while reproducing the lives of its family members. In order to bring out the labor force in a self-sufficient family economy into the labor market, it is necessary to bring it under the monetary economy under the above conditions. Because the penetration of the money economy progresses as an "irreversible cycle," family economies that are gradually drawn deeper into the money economy lose their self-sufficient livelihood. On the other hand, the family economy suffers from unstable employment conditions in the labor market and low agricultural prices, and is unable to obtain sufficient monetary income, losing its character as a self-sufficient economic unit. A similar theory of the dissolution of the subsistence family economy is reproduced in Meillassoux preface to Magasa (1978), published three years after "*femmes, greniers et capitaux.*"

The above is a sketch of the main points of the argument in Part 2 of Meillassoux (1975), with some comments. I have already given a critical examination of this argument elsewhere, and although there are some overlaps, (Yamazaki 2007) I would like to point out a few points below in line with the subject of this chapter.

The cause of low wages in developing countries that Meillassoux directly discusses is, in short, the transfer of value (rent in labor) from the non-capitalist subsistence family economy to the capitalist sector. Regardless of whether or not the value transferred in this way is called "rent in labor," the gist of this argument can be easily accepted even among Japanese agricultural economists. In the past debates about the Japanese style of low wages, one representative view was that workers' wages were generally low because agriculture covered part of the cost of reproducing the labor force. (Yamazaki 2014a)

Meillassoux suggests the following two forms of labor movement from non-capitalist sectors to capitalist sector that involve value transfer.

The first form (rural exodus), which can be inferred from Meillassoux's point that it corresponds to Marx's theory of primitive accumulation, is the movement of labor through the differentiation and disintegration of the peasant classes. Examples of this situation can easily be found in the past of developed capitalist economies, and it is an ongoing social phenomenon in developing countries in East Asia, where economic growth has been rapid in recent years.[122]

The second form (rotating migration), when expressed while emphasizing the contrast with the first form, does not involve differentiation or disintegration of the peasant classes. Therefore, it is a temporary movement of labor that assumes a return to the rural village of origin at some point in the future. Through this, value is transferred from other modes of production to the capitalist sector through low wages. This form is common in Africa after World War II, and includes not only labor movement within Africa but also movement to Europe. However, it seems to me that Meillassoux's discussion of this form has serious methodological problems that cannot be dismissed simply as "insufficient explanation." In the following, I would like to elaborate on this point, focusing on three key words; "type," "stage of development," and "community."

First, I would like to consider the question of how to understand the relationship between the second form and the first form. Will the difference between the two continue to exist as a type difference? Or, instead, can the relationship between the two be seen as a difference in developmental stages, and should the second form be understood as a transitional state that precedes the first form? Meillassoux's views on this point are not clear and seem to oscillate between these two views. Indeed, when Meillassoux discusses the two

---

122 Regarding Vietnam, see Yamazaki (2004, 2007). For Thailand, see Tasaka (1991). Kitahara (1985) discussed Southeast Asia in general, using Thailand and Indonesia as examples.

forms of geographical location (1st form = developed countries, 2nd form = Africa), there seems to be at least the beginnings of a perspective that sees the relationship between the two as a typological difference, although it is not so clear-cut. However, on the other hand, when Meillassoux argues that the self-sufficient family economy, which forms the basis of the second form of labor supply, will disintegrate due to the penetration of the monetary economy, he sees the relationship between the two as a difference in development stages. (It can be inferred that the first form of labor migration will probably take place after such demolition.) However, there has been no empirical examination of where the post-World War II situation in Africa that Meillassoux was facing lies within the series of trajectories drawn by the development stage theory (labor supply from subsistence family economy → its dismantling due to penetration of monetary economy). Therefore, it is difficult for the reader to judge from Meillassoux's text whether this logic, especially the logic of deconstruction in the latter part, is valid and consistent with the actual situation. Perhaps Meillassoux himself did not see this kind of disintegration process in reality. This is because, in contrast to Meillassoux (1975), which is a book on abstract logic that we have considered so far, Meillassoux (1964) is written mainly to describe the actual situation of the Guro people of Ivory Coast, in the latter, the strength of the subsistence economy and the traditional social relationships based on it is advocated as follows.

"The self-sufficiency economy in the food production sector is maintained, and around it, the same labor relations and social relations surrounding kinship and marriage persist." "The robustness of traditional societies is explained by their inclusion and preservation within the global economy as a necessary component. This is because it is currently the only form of social and economic organization that can satisfy the population's food needs and thereby supply the capitalist sector with people and products at low prices. From this point of view, traditional societies continued to play the role that colonization had imposed on them. That is, they have continued to prevent workers from fully integrating into the capitalist sector, and they have continued to exempt the capitalist sector from the costs that would be necessary if it were to provide for all the needs of the workers."

Thus, Meillassoux does not posit a historical relationship between the two forms. Or, to be more precise, since he does not seem to have raised this issue, he has not actually attempted to consider this point. It seems to me that behind this is the fact that he has not fully grasped the characteristics of the society that forms the background of the second form. This is because Meillassoux's *communauté domestique*, which forms the basis of the second form of labor supply and focuses on both material reproduction and life reproduction of a self-sufficient family economy, lacks the concept of "community" that characterizes the non-capitalist mode of production.[123] The question is why "traditional society" is so willing or able to resist the destructive forces of capitalist economy. In order to consider this problem, it is necessary to take the perspective that the second form exists while being connected to and being enveloped by the "community." Therefore, I will discuss this point in the next section.

4. Community-Based Labor Supply

When considering the issue of labor supply from non-capitalist production modes to capitalist sector in developing countries, it is necessary to take into account that there are cases where non-capitalist production modes still exist while being protected by communities. The theoretical standard when we

123 Otsuka (1955) characterizes pre-capitalist societies as ones in which communities exist extensively.

consider the problem of labor supply specific to capitalist society is the theory of the capitalist law of population, that is, the theory of labor supply through the increasing of organic composition of capital and the business cycle. On the other hand, there is the theory of labor supply in the process of dismantling the non-capitalist modes of production as the theory of primitive accumulation and the theory of decomposition of peasant classes. Unfortunately, however, it seems that we still do not have a theory of labor supply in a society where communities exist strongly. Therefore, there is a theoretical difficulty in considering this problem.

The first thing to confirm is that the purpose of economic activity in a "society with a strong presence of community = communal society" is significantly different from that of a capitalist society, which is the opposite (see previous chapter). In a communal society, the purpose of economic activity is to produce and secure use-values, mainly agricultural products, of quality and quantity that guarantee the survival and reproduction of individuals. This is a society that differs, at least on the surface, from a capitalist society in which each individual operates with the aim of maximizing economic value such as profit, rent, and wage, whether in the short or long term. That purpose, in turn, determines the specific nature of production relations and distribution methods within a communal society.

Communal ownership is the form of ownership of the natural production conditions represented by land, which is necessary to realize the purpose of such community economic activities. However, this does not mean that the above purpose can be achieved just by having this form of ownership. Communal ownership is a guarantee of the survival and reproduction of individuals in the input dimension of the primary means of production. However, on the other hand, it is also necessary to guarantee the survival and reproduction of individuals in terms of output. Even if the families that make up a community receive the necessary and sufficient amount of production means, represented by land, from the community, they may fail in the process of managing the production means. Or, the families may encounter a natural disaster, unexpected accident, or illness. In such cases, the families may not be able to produce enough consumer goods or inputs for the following year. In order for community members to share the consequences of risk in this broad sense, the community absorbs surplus products from those who have them, stores them, and then redistributes them to the poor. In this way, ex post adjustment of surpluses and deficiencies among members within the community must also be an important function of the community.

Therefore, in a society where such community functions remain strong, widening economic disparities and class differentiation among members are eliminated, and the sale of labor-power that would result from this is in principle non-existent. Therefore, if labor sales were to exist, they would need to be explained by a logic different from this mutual heterogeneity of members. They may be cases of general deprivation due to natural disasters that affect all members of the community. Otherwise, they may be due to the demand for money needed for "tax payments," etc. (Meillassoux 1975), which is not available in the subsistence economy that prevails in the community. Therefore, far from being the result of the dissolution of the community, the sale of labor-power in this case is, on the contrary, driven by the demand for money necessary to maintain the community.

However, this demand for money is not an essential condition for the reproduction of labor-power in normal times. In other words, the reproduction of labor-power by individuals in a community is basically carried out within the framework of a subsistence economy, supported by mutual aid within the community (stockpiling and redistribution). Therefore, the sale of labor-power from a community to the outside world do not have a permanent character.

This explains the existence of the "traditionalist behavior pattern of African workers who tend to stop selling their labor-power as soon as the target monetary amount is reached." (Akabane 1971) [124] In Sub-Saharan Africa, such communities still exist strongly today.[125]

What kind of influence, therefore, does a community have on the wage level of the labor it supplies? As we will see, the community exerts contradictory effects on the reproduction cost of labor-power and the trends in labor supply and its demand.

On the one hand, the existence of a community works in the direction of pushing down wage. The reason for this is along the lines of Meillassoux's argument that we saw earlier. The idea is that because the non-capitalist subsistence economy bears part of the cost of reproducing labor, the capitalist sector can pay lower wage for the labor supplied by the community. However, there is a point that must be emphasized as a special situation in societies where communities exist strongly. This is because the subsistence economy occupies a very high proportion of economic activity in such places, and a situation arises in which the subsistence economy bears a very large portion of the cost of reproducing the labor force. As a result, there is a tendency for the wage that the capitalist sector must pay to be further reduced. The following quotation from Volume 3 of "Capital" states that the price of agricultural products produced in a community can become so low that "the cost of production is not concerned," even if this is prompted by state demands. The reproduction cost of the labor supplied by the community is largely borne by

124 In this way, the labor force, which is basically reproduced through the subsistence economy within the community, is supplied to the capitalist sector driven by the temporary demand for cash. Because it is temporary, the labor force is supplied with the assumption that it will be returned to the original community. Elsewhere, I define this situation as "community-based labor supply." (Yamazaki 2007)

125 Previous chapter, Section 6. Yamazaki (2007).

the community's self-sufficient economy, so wage can be low, with no regard for the reproduction cost of the labor force.

"And, on the other hand, there were the land holdings of Russian and Indian communist communities which had a sell a portion of their produce, and a constantly increasing one at that, for the purpose of obtaining money for taxes wrung from them — frequently by means of torture — by a ruthless and despotic state. These products were sold without regard to price of production, they were sold at the price which the dealer offered, because the peasant perforce needed money without fail when taxes became due." (Marx 1894) [126]

This situation, in which the self-sufficiency economy is responsible for the majority of the reproductive cost of labor, continue to persist stubbornly, supported by communal mutual aid. This is one effect that communities have on wage level.

However, secondly, the trend of lower wage as pointed out in the first point does not actually materialize. Therefore, it remains a latent tendency. In other words, wage is actually higher in society where community exists strongly than in society where it is not. The reason is as follows.

Due to the circumstances mentioned above, in society where community exists strongly, the portion of the labor reproduction cost that should be borne by the capitalist sector can be made cheaper. However, for this to actually materialize in the form of low wage, surplus-population pressure in the labor market is necessary. However, it goes without saying that in society where community exists strongly, the creation of a relative surplus-population due to the increasing of organic composition of capital in response to capital accumulation is not observed. Furthermore, the supply pressure on the labor market through the differentiation and disintegration of direct producers,

126 Just to be sure, this text was written by Engels, the editor.

centered on the peasant class, has not yet begun to take effect in earnest. In that sense, "the separation of the laborer from the conditions of labor and their root, the soil, does not yet exist," the labor market is always in short supply, and "the laws of the supply and demand of labor falls to pieces." (Marx 1867) This is Marx's description of colonies where the land belonged to the people, but a similar situation can be seen in a society where community exists strongly. In this way, in a communal society, the aforementioned tendency toward lower wage remains latent, blocked by the effects of labor supply and its demand. Actual wage will be higher than in other developing countries where differentiation and disintegration of direct producers have already begun. In fact, Hirano (2005, 2013) points out the existence of "high wage in the formal sector manufacturing industry" in Sub-Saharan Africa, compared to East Asia, which has a comparative advantage in cheap labor.

However, a community is not necessarily a federation of homogeneous members. There may be cases in which the production relations within a community are dominated by class relations between landlords and tenants, and it may even be said that this is more common. However, since class relations within a community are based on the functions of stockpiling and redistribution, the communal mechanism that protects its members and prevents their differentiation and disintegration still functions strictly there. (See previous chapter.)

5. Conclusion

The problem posed at the beginning of this chapter was to organize theoretical points regarding the causes of national disparities in labor wage (relatively low wage in developing country). The following points became clear through the discussion in this chapter.

First, because there are differences in the intensity and productivity of national labor between latecomer and first-mover countries in capitalist societies, the law of value is "modified" when applied internationally. As a result, the value (price) of workers' daily necessities generally tends to be lower in latecomer country and higher in first-mover country. This difference in price level between countries are the cause of relatively low wage in latecomer country, assuming other conditions such as the standard of living of workers are held constant.

Second, it goes without saying that agricultural products are important among the daily necessities of workers. The price of agricultural products can be lower when agriculture is carried out by peasants than when it is carried out by the capitalist sector. This is because the former allows for distress sales. Furthermore, even when agriculture is equally carried out by peasants, the degree of establishment of C and V categories is higher in developed countries than in developing countries, and therefore the degree of distress sales is more pronounced in developing countries. Therefore, the price of agricultural products can be lower in developing countries from this point of view as well. However, if landowners absorb this difference as rent, the price of agricultural products will rise by that amount.

Third, the non-capitalist mode of production that currently exists widely within developing society provides low wage to capitalist sector by becoming a source of labor supply for the capitalist sector through its maintenance, differentiation, and decomposition. In other words, the non-capitalist mode of production can supply cheap labor to the capitalist sector by bearing part of the labor reproduction costs. And if such labor supply occurs through the rapid differentiation and disintegration of non-capitalist modes of production, this will put surplus-population pressure on the labor market, resulting in low wage.

Fourth, however, when non-capitalist modes of production exist together with strong community, and the latter hold strong influence in society, the degree to which labor-power reproduction is generally carried out by the subsistence economy is large. Therefore, there is a potential for the capitalist sector to extract labor from non-capitalist modes of production at very low wage. On the other hand, however, because a community suppresses differentiation and disintegration among its members through its internal mutual aid function, it becomes difficult for surplus-population pressure to form in the labor market. Due to the latter situation, the aforementioned potential is not reflected in the actual wage realized through the supply and demand relationship in the labor market. Actual wage is generally higher than wage formed under the circumstances described in "third" above.

One issue that is still missing from the above explanation is the issue of the disparity in the range of daily necessities that determines the value of labor-power that exists between developed and developing countries. In other words, it is a problem of disparities in real wage or "standards of living." However, we can accept that the fact is that real wage is relatively high in developed country and relatively low in developing country, although it may take some twists and turns to explain this. Therefore, it would be permissible to add this to the list of factors that lead to relatively low wage in developing country.[127] It is also necessary to consider what kind of revisions the theory of national wage disparity developed in this chapter will undergo when the three

127 Marx also discusses the relatively high real wage earned by workers in developed country, although he is limited to "workers in manufacturing industries" as follows: "The more productive a country is in the world market compared to other countries, the higher its labor wage compared to other countries. In Britain, not only nominal wage but also real wage is higher than on the continent. Workers are eating more meat and satisfying more desires. However, this applies only to workers in manufacturing industries, not to agricultural workers. However, the wage of workers in manufacturing industries are not higher in proportion to the productivity of British workers." (Marx 1862-1863)

recent factors mentioned in Notes 100 and 117 are incorporated into the discussion; ① existence of monopoly prices, ② high productivity of agriculture in developed countries of the New World, ③ industrialization of developing countries.

Chapter 6 : Primitive Accumulation in "Peripheral Regions" : Southeast Asian Model and West African Model

1. Theme

According to Wallerstein (1974), modern capitalist society was born in the environment of the self-sufficiency zone with the wide-area division of labor of the "European world economy," (including the New World) which was established during the "long 16th century" from the late 15th century to the first half of the 17th century. He also argues that the formation of the "European world economy" was at the same time the establishment of a three-tier regional structure; "core," "semi-periphery," and "periphery." At that time, *Gutsherrschaft* was established in Eastern Europe, and a forced labor system for natives called encomienda was established in the Spanish New World, which was the formation of a "periphery" that corresponded to the formation of a commercial and industrial center in northwestern Europe. The establishment of a "core" was only possible through the formation of a "periphery" that provided raw materials, food, and labor. At the same time, the Mediterranean region, which was once the most developed region, became "semi-peripheral." [128]

The world economy since then has become more extensive than the "European world economy" of the 16th century. The mid-19th century was the period of the British system (Mochida 1996), with Britain as the only industrial country and the rest as agricultural countries. The period from the 1870s to the end of the 19th century was a period of transportation revolution centered on railroads and steamships, and a parallel transition to imperialism

128 The concepts of "core" and "periphery" are from Amin (1972, 1973), Amin et al. (1982).

(colonial system and competition between multiple industrialized countries). Furthermore, since the 20th century, the world economy has literally become global. However, even then, the fact of "a mutually complementary bifurcation into the core and periphery within the world economy" is recognized.

As is often said, Wallerstein's approach tends to lead to pessimism regarding developing economies, which have been forced to play a "peripheral" role. However, in reality, there are cases in which remarkable economic development blossoms from the "periphery." [129] On the one hand, Fröbel's theory of new international division of labor depicts the multilayered structure of the world economy in the vein of Wallerstein, but on the other hand, it dispels Wallerstein's pessimism and preaches the dynamism of the development of developing countries' economies. (Fröbel et al. 1980) The theory of new international division of labor states that the capitalistization of "peripheral" economies cannot be viewed solely as an economically colonial and dependent transformation process in which they are positioned as sources of industrial raw materials and food for "core" countries. On the contrary, this theory discusses the development of developing countries' economies. This development is viewed in terms of capital accumulation that progressed on a global scale after World War II. This theory also depicts this development as a combination of foreign direct investment from the "core" and internal factors in the "periphery" (formation of a global industrial reserve army).

What is, on the one hand, a feature of this theory of new international division of labor, and at the same time, what can be seen as its problem, is that the picture it depicts of capital accumulation is a little excessive in its assertion of "universality" among developing countries. On the other hand, there is a lack of interest in the regional disparity in economic development that actually exist

129 "The export of capital influences and greatly accelerates the development of capitalism in those countries to which it is exported." (Lenin 1917)

between developing countries. (Yamazaki 2007) In fact, one of the notable features of developing economies since the 1980s has been their polarization among countries, marked by the contrast between high growth and stagnation. If this is the case, we must now identify the problems inherent in the theory of new international division of labor.

Here, I would like to point out two general problems regarding the theory of new international division of labor. One is the lack of analysis of the agricultural structure of the "periphery." Analysis of internal factors in the "periphery" cannot be lumped together with the aforementioned concept of the formation of an industrial reserve army on a global scale. The diversity and regional characteristics of the formation process of the industrial reserve army that actually exists among developing countries must be discussed (including cases where the industrial reserve army is not fully formed). This is all about discussing the primitive accumulation process and the diversity and regional characteristics of its existence in each country. There are violent aspects and economic aspects to primitive accumulation, and the latter mainly involves the differentiation and disintegration of the peasant class. It will be necessary to conduct a comparative study of these two points among developing countries, and then proceed to categorizing the primitive accumulation in "periphery" based on this comparative study. This is a more in-depth consideration of Lenin's (1917) statement that "the conditions for industrial development have been created" as a factor that provides "the possibility of capital export." The second problem with the theory of new international division of labor is the following: This theory does not pay sufficient attention to the diversity of forms in which "core" capital attracts and utilizes the labor-power flowing from "peripheral" pre-capitalist elements. This diversity in forms of attraction and use is the diversity of external relations between "peripheral" pre-capitalist elements as sources of labor-power and capital in the "core" as demanders of labor-power. This can also be described as the diversity of relationships of "articulation" between the former and the latter through the labor market.[130] Furthermore, since this is rooted in and determined by the nature of primitive accumulation in the "periphery" and its national differences, this is also related to internal factors in the "periphery." Therefore, the second problem is closely intertwined with the first problem.

With this awareness of the problem, the task of this chapter is to discuss two contrasting forms of primitive accumulation in developing countries considered by Amin and Wallerstein as "peripheral regions" (Amin, 1972; Amin, 1973; Wallerstein, 1974; Amin et al., 1982) ; the Southeast Asian model and the Sub-Saharan African model. This is positioned as an initial task, while looking ahead to a systematic and comprehensive typology of the "peripheral" primitive accumulation process. In order to be able in the future to propose a systematic and global typology of the process of primitive accumulation in the "peripheral regions," I will try to present here two extreme forms. In recent years, one of these regions has stagnated while the other is booming economically. If Africa joined the capitalist world earlier than Asia,[131] real economic development has long remained the prerogative of some of the countries of northern Africa and South Africa. It took until the end of the 19th century for economic "development" to affect the interior of the continent, but the struggles between the great European colonial powers rapidly intensified at this time and the economic "stagnation" that set in after the World War I is still noticeable. In sub-Saharan Africa, the average annual growth rates of per capita GDP were 0% (1970-90) and 1.8% (1990-2009). (UN) World War I, which broke out in 1914, was initially limited to Europe. However, the British

130 Regarding the concept of "articulation," see Mochizuki (1981a).

131 According to Wallerstein (1974), Asia was not part of the "European world economy" that was established in the 16th century.

and French Allied Forces invaded Germany's colonies, and, one year after the start of the war, the war spread to the African continent. At that time, European powers forced residents to leave and conscripted them as labor and soldiers, resulting in the fundamental destruction of people's lives. On the other hand, Southeast Asia has completely emerged from the stagnation in which it was plunged. The economic development, described as an "East Asian miracle," (World Bank 1993) which Southeast Asia has achieved in recent years together with the rise of all the countries of East Asia is remarkable (average rates of growth for East Asia and the Pacific of 5.4% and 7.2% over the same periods). Southeast Asia and sub-Saharan Africa therefore both suffered a long economic stagnation, but there are great differences in their development, particularly from the 1980s. As we will see in the following sections, this contrast can, in our view, be explained by distinguishing between two forms of primitive accumulation in the "peripheral regions." In the 4th section, I will discuss the differences in rural land tenure systems, which seem to me to be at the origin of the differences in the forms of primitive accumulation.

2. Definition of Primitive Accumulation in General

However, before proceeding to develop the opposite types of primitive accumulation in the "periphery," it is necessary to confirm the general definition of primitive accumulation.

Primitive accumulation is a process in which direct producers, mainly peasants, who were previously tied to the means of production, are historically separated from the means of production (land). The proletariat that was created through this process was drawn into industrial capital in search of a way to live. (Marx 1867) Through these events, social control by industrial capital was established. Moreover, the primitive accumulation is a one-time social phenomenon, regardless of the length of the period. However, this historical process of creating the relationship between capital and wage labor did not take place all at once on a global scale at the same time. Primitive accumulation began in Western Europe, centered around Britain, and later spread to surrounding areas. At that time, the relationship between capital and wage labor was summarized on a country-by-country basis, and capitalist societies were born regionally in so-and-so countries. Therefore, for example, "British capitalism" and "French capitalism" appeared with each country's name as an adjective. If we think of things this way, primitive accumulation is a national process. However, if we look at things differently and consider the history of global capitalist society as a one-time process that started with the primitive accumulation in Britain, then primitive accumulation has not yet been accomplished even in the 21st century, as evidenced by the fact that it is either ongoing or undeveloped in today's developing countries. The validity of Luxemburg's (1913) long-term vision, which holds that a capitalist society is compatible only with the continuation of primitive accumulation, is not addressed here. However, the reality is that the past development process of global capitalist society has always coexisted with primitive accumulation, or has had primitive accumulation as one of its components.

Furthermore, based on the recognition of the multilayered structure of the world economy, primitive accumulation consists of at least two mutually complementary processes: a "core" process and a "periphery" process. (Fujise 1985) Furthermore, in each period, the "periphery" maintained its character as a source of raw materials and food for the "core," while promoting industrialization by importing production-goods and capital from the "core." Such industrialization of the "periphery" naturally presupposes the existence of the primitive accumulation process in it. Even today, the "core" and "peripheral" processes of primitive accumulation are ongoing as two contradictory yet complementary processes of the primitive accumulation of global capitalist society. In the "core" the primitive accumulation has already been accomplished or is nearly accomplished, but in the "periphery" it is currently underway or is a future process depending on the location.

3. Two Opposite Forms of "Peripheral" Primitive Accumulation—Southeast Asia Type and West African Type

I approached the analysis of the differences in development between South-East Asia and sub-Saharan Africa from the two systems that are the "East-Asian complex" and the "Europe/Sub-Saharan Africa complex," each being made up of a developed "core" (Japan, Europe) and an economically underdeveloped "peripheral region" (South-East Asia, sub-Saharan Africa). (Yamazaki 2007) Furthermore, I analyzed the agricultural structure, the main content of which is the differentiation and decomposition of the peasant class, with the aim of discussing the supply mechanism of labor derived from each "peripheral" pre-capitalist element to the capitalist sector. Consider first the East Asian complex. From the 1980s, the drying up of the Japanese labor force of rural origin led to an increase in direct foreign investment by Japanese industrial companies in the direction of the "peripheral" regions of Southeast Asia.[132] The differentiation of peasant class that had accompanied the Green Revolution in the "peripheral" regions had indeed created a reserve of cheap labor, which attracted foreign direct investment from the "core" of Japan. This perfectly follows the pattern of development described by studies of new

132 Maswood (2018) argues that protectionist policies adopted in the United States in the 1980s to save domestic industries from competition with Japanese companies triggered the international expansion of Japanese manufacturing (formation of an internationally divided manufacturing process network). Furthermore, he argues that the global economy was realized when European and American manufacturing industries learned and imitated the Japanese company's method of using affiliated companies.

divisions of labor on an international scale. (Fröbel et al., 1980) Yamazaki (2004) described the process of differentiation and decomposition of the peasant class, focusing on the Mekong Delta during the Doi Moi period. By referring to several previous studies, it becomes clear that the progress of differentiation and decomposition of the peasant class due to the "Green Revolution" is a social phenomenon that covers a wide area of Southeast Asia. (Kitahara 1985, Tasaka 1991) It goes without saying that the process of differentiation and disintegration of the peasant class presupposes the existence of a farmland market, and the latter presupposes the existence of the idea of private rights to farmland.

However, although agricultural research in Southeast Asia is flourishing these days, there are surprisingly few scholars who deal with the "classical" theme of differentiation and decomposition of the peasant class. Furthermore, it seems difficult to say that the significance of such research is fully understood today. In fact, the significance of research on the differentiation and decomposition of peasant classes in Southeast Asia may not be fully understood unless it is discussed as a comparative agricultural structure theory with sub-Saharan Africa in mind. Otherwise, the "unusual character" of the phenomena observed in Southeast Asia would not be noticeable.

In contrast, in many parts of sub-Saharan Africa, the dominant land system of community ownership has prevented the development of active rural land markets, and therefore of strong differentiation dynamics of the peasant class. Other factors reinforce this phenomenon; the lack of competition in terms of productivity, and the lack of disparity in terms of profitability between farming families. Thus, the land system and rural social structures of sub-Saharan Africa offer a striking contrast to the active social differentiation that take place in Southeast Asia.

On the other hand, there is a fear that the above description of rural areas in sub-Saharan Africa may actually be a way of expressing it from a too modern perspective. Traditionally, relationships between individuals in agricultural village in sub-Saharan Africa are not based on "competition" but on "symbiosis" and "coexistence," at least within community. The structure of society is therefore fundamentally different from that based on competition found in Southeast Asia. In other words, the economic activity of village in sub-Saharan Africa is more linked to a "use-value" of the product than to an "exchange-value" of the product. (Sahlins, 1972) Yamazaki (2007) discusses the above points with reference to the Inner Niger Delta, but what can be generalized across the regional spread of sub-Saharan Africa is that some previous studies have described similar situations. (Akabane 1971, Sahlins 1972, Sugimura 2004) Thus, the farmers who have to leave their land, that is to say the laborers resulting from the differentiation of peasant class, do not emerge as the own social class. On the contrary, the seasonal salaried activities that can be observed in the cities of sub-Saharan Africa aim to earn the money necessary to maintain agriculture within the village. In short, one of the main objectives of emigration is to prevent the disintegration of subsistence farm economies. Labor migration from villages to cities and wage labor in cities in sub-Saharan Africa therefore presuppose a possible subsequent return to the village. (Meillassoux 1975)

In this context, sub-Saharan Africa has failed, and still fails, to truly attract foreign direct investment from the "core" for its industrialization. The recourse to cheap labor from the agricultural regions of the developing countries by the capital of the European "core" therefore takes the form of migration of labor which will settle in the "core." In the past, such migration first took place as slave trade. These days, it takes the form of international migration. It may sound a little paradoxical to make the laborers who come from "peripheral" rural areas with intention of returning to their villages, settle in the "core" and to use them as immigrants. However, if they are left in sub-Saharan Africa, they will return to the rural area, and they cannot be fully subsumed under capital. In order to fully incorporate them into capital relations, it is necessary to geographically and institutionally isolate them from their place of origin. To achieve this, immigrants need to be permanently residing in the "core." Yamazaki (2007) discusses immigration trends in France, the former colonial power in the Inner Niger Delta. However, this relationship between the "core" and the "periphery" regarding the use of labor can be generalized as the relationship between sub-Saharan Africa and Western Europe.[133]

4. Land Tenure System and Agricultural Productivity

1) Rules Governing the Land System

The differences between land tenure systems in Southeast Asia and sub-Saharan Africa (individual land ownership on the one hand and community land ownership on the other), themselves products of history, are probably origin of the opposing patterns of primitive accumulation in the two "peripheral regions" studied. But what is the origin of this difference between land tenure systems? Let us approach this question from the point of view of agricultural productivity, based on the data collected by the present author (surveys of farmers and local organizations, statistical data, official documents) in the Mekong Delta (Vietnam) and in the Inner Niger Delta (Mali) between the 1990s and the 2000s. (Yamazaki 2007)

Different methods of wet rice cultivation have developed in the Mekong Delta, depending on the climate and natural conditions of each region, including rainfall, soil quality, salinity, floods and tides. If the development of these

133 However, as pointed out in that book, this does not mean that former colonies in sub-Saharan Africa are the main source of immigrants to France.

different rice growing methods presupposed the existence of hydraulic developments undertaken by the colonial administration and then by the State of Vietnam, it is also the result of the adaptation of farmers to the natural conditions of each region. Investments by farmers themselves in the construction of bunds and end channels have played an important role in the process of adoption of modern rice cultivars developed from the 1970s by research organizations.

If we compare the process of increasing rice productivity and the level it has reached in the Mekong Delta (10 t/ha/year) with those of rice growing in the Inner Niger Delta (controlled flooding; 1-2 t/ha/year and Office du Niger; 6-7t/ha/year), the differences are significant. This difference is also due to the fact that the fairly widespread method of wet rice cultivation in the Inner Niger Delta is the traditional method of floating rice, which is used in fields where the water level is not controlled (freed flooding). [134] Even today, traditional rice cultivation is often carried out under conditions of low level of water control (controlled flooding). (Barbier et al., 2011, Jamin et al. 2011) In addition, the construction of water supply infrastructure necessary for the establishment of irrigated rice cultivation with water management, which can be observed today in the upstream part of the Inner Niger Delta (at the Office du Niger), almost always depends on financial and technological assistance from foreign countries. (Coulibaly et al. 2006, Bélières et al. 2011, Kuper 2011, Adamczewski et al. 2011, Kuper 2011) It is still rare for local farmers to take the initiative for new developments, carry them out and invest themselves, even if the situation has started to change since the late 2000s. (Brondeau 2011, Barbier et al. 2011)

The differences between the two deltas, in terms of historical processes of

134 For information on various rice cultivation methods in the Inner Niger Delta, see Yamazaki (2007).

increasing swamp rice productivity and current productivity levels, can be characterized as a real "disparity" and probably reflects the difference between the dominant land tenure systems between them.

In the Mekong Delta, the Nguyen dynasty fixed the land tenure system at the beginning of the 19th century. (Brocheux 1995) It even seems that the concept of land ownership was already widespread before this time. However, the land system in place in the Mekong Delta since the establishment of the socialist regime in the mid-1970s is a collectivist land system. But since the land laws of 1988 and 1993, farmers have obtained rights, equivalent to individual property rights, over the land they cultivate. A market for the sale and rental of agricultural land is thus developing. (Cho et al. 2001, Yamazaki 2004)

In contrast, as in many other agricultural areas of sub-Saharan Africa, the community land management system is traditional in the Inner Niger delta and continues to dominate, except in the Office du Niger area.[135] It is considered that the land belongs to the community and that its management is entrusted to the "master of the land," or land chief, by all the members of the village. This "master of the land" embodies land ownership by the community. The community land is distributed by him, and this system allows the farmers to have the usufruct. In addition, farmers can pass on to their descendants the

135 Immediately after gaining independence in 1960, the Mali government advocated a socialist policy modeled on China and declared the entire country state-owned. However, under the ensuing economic crisis, Mali applied to return to the franc zone. Furthermore, the military government that was established as a result of a coup in 1968 attempted to significantly revise the socialist path of the previous government. Despite these political upheavals, in rural areas, state ownership of land has been maintained to this day. However, this system of state ownership of land is actually something that hardly ever comes to the surface of social life. Even today, what governs the reality of rural life is the customary ownership rights and usufruct rights over land, as described in the text. Also, due to Islamic tradition, land taxes are not imposed by the state.

land that has been allocated to them. On the other hand, they are prohibited from selling the land, mortgaging the land, lending the land against remuneration. Moreover, if the land has not been cultivated for a long period (of the order of several years), or if the beneficiary of the land does not manage it properly, he must return it to the "master of the land." (Cissé 1982, Schmitz 1986, Barrière et al. 2002, De Noray 2003).

To sum up, it seems that the concept of individual land ownership developed quite early in the history of the Mekong Delta, which allowed long-term agricultural investments. In contrast, in the Inner Niger Delta, community land ownership has hampered sustainable agricultural investments by landholders.[136]

In addition, in the Mekong Valley, there is a land market within the framework of which competition between farms in terms of productivity and the differentiation of peasant social class develop. On the other hand, in the Inner Niger Delta, community ownership does not encourage the development of competition between farms in terms of productivity, nor the differentiation of peasant social class. On the contrary, the primary goal of relationships between individuals is to allow them to live together, even if it is in a very hierarchical society. (Gafsi et al. 2007) [137]

However, this reasoning that the disparity between the two deltas in terms of investments in land development by farmers is due to the difference between their land systems, leaves aside the fundamental notion of

136 It is a widely held view that the cause of Africa's agricultural stagnation is its communal land ownership. (Yoshida 1999)

137 Sugimura (2004) analyzes the process by which the "rich-poor" relationship is created in the Qumu society of Zaire. This is not a process of establishing "relationships of domination" in which differences in the means of production widen. Even today, this process is supported by the "relationship of distribution," in which wealth always flows from the rich to the poor, and by the unique wealth structure that originates from prestige goods.

productivity. In general, when the development of agricultural productivity reaches a high level, the concept of private ownership of land gradually emerges. Indeed, the person who invests to clear the land, increase its fertility or irrigate it, must be certain of being able not only to recover his investment but also to benefit from it. In this case, it becomes essential to establish, in custom or in law, a relatively strong individual land right.[138] In this view, the concept of individual land ownership becomes the product of a certain degree of development of agricultural productivity.

In this regard, Iliffe (1983) argues that if the concept of individual right to land has not developed in Africa, it is because there is ample space in relation to the population and the land has not become a scarce good. Indeed, the then densities in the regions studied here were 18 inhabitants/km2 (in 1998) in the Mopti region where the Inner Niger Delta is located, against 422 inhabitants/km2 (in 2002) in the Mekong Delta. In addition, according to a survey by the local Malian administration, the proportion of cultivable land actually cultivated does not exceed 11% in the Inner Niger Delta. (DIRASSET 1994) When the development of agricultural productivity reaches a level requiring investments in land, it generally becomes necessary to introduce the concept of individual land rights into society, regardless of population density. Indeed, when in the early 19th century the Nguyen dynasty established the individual land system in the Mekong Delta, large areas of the delta were still unexplored.

138 Sakurai (2005) examined existing literature and clarified the following points. In Sub-Saharan Africa, medium- to long-term investments in land are made even under "customary land systems," and these investments tend to strengthen land ownership. These propositions are verified through correlation analysis using his own survey data in Côte d'Ivoire. Furthermore, according to Shimada (1975), who refers to Netting's theory, in the "terraced cultivation area" of northern Nigeria, when up-front labor investment such as terracing, dumping large amounts of compost, and caring for tree crops increases, peasants begin to insist on private rights over land.

Furthermore, Akabane (1971) argues that the "individualization of land ownership" caused by "population growth" does not result in the dissolution of the "community" because it is not the result of farmers' rights arising from increased productivity. Using the example of the transfer of perennial trees between individuals, he argues that the principle of "labor investment" operates as the basis of private rights to land. However, on the other hand, Akabane also writes that "community land ownership" and the transformation of "an old human type with a traditionalist mental structure" will enable the development of "social productive forces." Regarding the relationship between productive forces and production-relations/superstructures, he does not only regard the former as determining the latter, but also focuses on the impact of the latter on the former. From Akabane's perspective, the present author's perspective above places far more emphasis on the fundamental nature of productivity. However, I think Akabane's problem is that the relationship between agricultural productivity and production relations (land system) is not discussed organically and in an unmediated way. In my opinion, consideration of the differentiation and disintegration of the peasant class is the main link that connects the two.

2) Natural Conditions and "Models" of Agricultural Productivity

If individual land rights remain poorly developed in the Inner Niger Delta, it is because the form of wet rice cultivation that still dominates there does not generally require large investments. The level of productivity requiring the emergence of individual land rights has not yet been reached. The following problem then arises: What is at the origin of the characteristics (which can be described as a "model") of these production capacities, which cannot be reduced to the difference in development degree between wet rice cultivation in the Inner Niger Delta and the Mekong Delta? I will explore this point below, in relation to climatic, natural conditions and historical processes.

First of all, the climate of the Inner Niger Delta is semi-arid (rainfall less than 500 mm/year). The initial investments related to land clearing are therefore small. The conditions are very different in the Mekong Delta (monsoon; average annual rainfall: 1,500 mm/year), where it was necessary to cut down the cajuputs (*Melaleuca cajuputi*) and the mangroves (with Myrsinaceae, Rhizophoraceae, Verbenaceae, etc.) which were invading it and where it was necessary to dig evacuation channels where the water flowed with difficulty. In other words, considerable initial investments were needed to clear it. We also find similar conditions on the African coast of the "Southern Rivers" (Casamance, Guinea Bissau, Guinea), where paddy rice cultivation can be seen on the cleared land of the mangrove forest. (Cormier-Salem 1999)

Secondly, West African aquatic rice cultivation was initially practiced by following the rhythm of the natural cycles of the floods of the Niger River. Water control was still weak before the Europeans introduced large hydro-agricultural developments in the 20th century. In East Asia, on the other hand, irrigation has a very old history. (Watanabe T. 1977, Iinuma 1970, Yanagida 1961, Vidal de la Blache 1922) In the climatic and natural conditions of the Inner Niger Delta, the bases for developing this kind of agricultural know-how were absent. Yet between the 8th and 16th centuries, the kingdoms of Ghana, Mali and Songhay flourished and then declined in the Inner Niger Delta. It is likely that these kingdoms had not developed large hydro-agricultural schemes. (Gallais 1984) These kingdoms were not based on agricultural production, but rather on the trade of gold and salt.[139] On this point, the history of the central

139 Kawata (1981), Miyamoto et al. (1997). However, there are objections to the way this sentence describes the foundations of existence of the West African kingdoms. Takezawa (1984) states that rice cultivation was one of the economic and social foundations of both the Kingdoms of Mali and Songhay.

delta of Niger is very different from that of eastern Asia where states have developed since antiquity and the Middle Ages, a state system called "Asian despotism" (Montesquieu 1748, Marx 1858, Wittfogel 1957) based on the management of large-scale irrigation facilities, such as the Nguyen dynasty in the Mekong Delta. Thus, natural conditions in West Africa have prompted neither farmers to increase their productivity in terms of water management, nor the State to invest in hydro-agricultural development.

This lack of water management ideas in the Inner Niger Delta seems similar to the situation in Egypt, which Watsuji (1935) described as "a strange dual character of dryness and wetness." Watsuji, who said that Egypt's "dry and wet" climate and natural conditions have given rise to a passive attitude of people toward the rising waters of the Nile, would probably say the same about the Inner Niger Delta. However, in the Mekong Delta, where the climate and natural conditions are "humid" like the monsoons, people have been acting more proactively toward rivers, as mentioned earlier. Therefore, it must be said that it is impossible to derive flood control methods straight from only the climate and natural conditions, as Watsuji did. Commenting on Watsuji's theory of climate, Inoue (1979) rightly argues as follows: "It may be a clue to understanding ethnicity in art, religion, and philosophy. However, it does not take into consideration the economic aspects of humans, especially the place of production. In other words, it is inadequate because it does not take into account both the active and passive relationships that exist between humans and nature." It is true that the historical process of the mutual relationship between humans and nature that has developed in the field of paddy cultivation is largely framed by climate and natural conditions. However, the former needs to be discussed separately from the latter. In other words, it is

---

Furthermore, Takezawa et al. (2005) emphasizes the relationship between fonio cultivation, iron manufacturing, and the establishment of Ghana and Mali.

necessary to pay close attention to the aspect of active efforts by humans. Moreover, Iinuma (1970) states: "The climate is a natural framework that cannot be changed by human power. However, how to use it depends on the subjective conditions on the human side (simply put, the state of capital and labor)." From this point on, he proposes the idea of "dynamic theory of climate." Additionally, Febvre (1922) opposed "determinism" and proposed the concepts of "humanity as a natural cause" and "a humanized earth." Both can be said to be outstanding.

Returning to the discussion on the "model" of rice productivity, third, aquatic rice cultivation in West Africa has the particularity of being linked to livestock. (Dugué 2003) [140] Indeed, the paddy fields, after the harvest, serve as pasture for cattle, sheep, goats, horses, donkeys and camels. The passage of these animals destroys the rice fields ridges, which is a problem for the rice-growing developments of the Office du Niger. (Bonneval et al. 2002) This is one of the obstacles to investment in rice development. This historical characteristic of West African rice cultivation constitutes a great difference

140 On the other hand, according to Nakao (1966), the cultivation method in the savannah farming culture area, including the Inner Niger Delta, developed as hand-hoe farming, and the use of livestock dung did not make much progress. Neither the field-grass method nor the two-field method was created there. He says: "One of its developmental weaknesses lies in its lack of livestock." The pastoralism that plays an important role in West Africa today was introduced by the Caucasian people. However, because it was at a very low stage of development, the spread of pastoralism had a rather suppressive effect on the development of agricultural culture in West Africa. (Nakao 1969) There is a similar description in Febvre (1922), which further describes the significance of livestock ownership practiced in West Africa as follows: "Satisfied with possession, saving, a portion of which is used carefully and only in times of extreme necessity; having a herd of livestock that has no value other than the value of dormant capital." Keller (1919) also notes that "originally only the Hamitic people owned cattle; later on, talented tribes among the Negro peoples similarly introduced them into South and West Africa." Regarding the low milk production per individual cow in Africa, see Sasaki (2000).

with that of East Asia, where the rice fields are not used as pasture after harvest. (Kumashiro 1969) The rice-growing "model" of the Inner Niger Delta has been shaped by semi-desert climatic, uncontrolled flood, and livestock associated with agriculture, namely by natural conditions and history. On the other hand, in Southeast Asia, it is the initial dense vegetation (the jungle and the mangrove), the management of water and the absence of pasture,[141] which forged the rice "model."

Although it is outside the scope of this chapter, let us take a look at rice cultivation in East Asia, which covers a slightly wider area than Southeast Asia. This observation shows that private rights to land have historically been nurtured within pre-capitalist communities. For example, in Japan, the following process occurred.

First, under the *Ritsuryo* system (7th-10th century), all land and people belonged to the state (public land and public citizen). Every six years, the government created household registers and allocated land to all the people, but when a farmer died, the land was returned to the state. According to the concept of "public land and public citizen," cleared land should also belong to the state, but the powerful classes of the time, such as aristocrats, shrines and temples, put pressure on the government to separate cleared land from the public land. Permanent private land exempted from paying tribute to the state was recognized as a manor (743; law that makes clearing permanent private property). Those who were dominant among the citizens devoted themselves to clearing by imitating the methods of the aristocrats, shrines and temples,[142]

141 However, in Southeast Asia as well, small numbers of cows and water buffaloes have been kept to provide labor and excreta (a source of compost) for the smooth running of arable agriculture. Their main feeds are wild grasses and cultivated agricultural by-products including rice straw. In addition, small numbers of pigs, chickens, and domestic ducks have been raised to help farmers make a living by using various resources around them as feed. (Sasaki 2000)

142 A literary work based on this topic is "Sansho Dayu." (Mori 1915)

and became a class called *myousyu* and *tato*.[143]

Under the Kamakura shogunate, manors became fiefdoms, and this led to the establishment of a feudal system based on private ownership of land. Later, the Tokugawa shogunate sealed off the so-called 300 feudal lords, excluding the lands of the shogunate, the emperor, temples and shrines. At that time, despite the shogunate's permanent ban on the sale and purchase of farmland and restrictions on farmland movement under the so-called limited land law, sales of farmland actually took place.[144] As a result, social stratification in rural areas has progressed into landowners (*goshi, kusawake, takamochi*), independent farmers (*neoi,* b*unzuke-hyakusho, kehou-byakusho*), tenant farmers, and half-tenants and half-workers (*mizunomi-byakusho*). Naturally, such agricultural land sales are based on the social existence of the concept of private land ownership. However, it is said that the dominant tendency in rural areas during the feudal era was the artificial suppression of class differentiation, and rather the universalization of poverty. (Noro 1930)

---

143 "The free clearing of land by villagers in the Nara period should also be understood from the characteristics of village cohesion during this period. There were no communal restrictions from the village side on the 'farmers' land' that were widely established then. The private ownership of clearing farmers over the land was permanent." "This weakness in village cohesion led to the widespread division of village communal land among dominant households and the widespread ownership of land, which led to the separation of the classes of *myousyu* and *tato*, who were based on land ownership. Due to the lack of village community regulation, the form of land division by dominant households is thought to have been the so-called pre-occupation system." (Ishimoda 1946)

144 "According to the 'Collection of National Civil Customs', in order to avoid violating the prohibition on the permanent sale and purchase of fields during the Tokugawa era, people called farmland purchase and sale '*yuisyoyuzuri*' or '*yoshiyuzuri*' and exchanged and received compensation in the name of key money or price for barrel sake and appetizers. Transfers of land in this manner were carried out throughout the country." (Nakata 1923)

5. Outlook

At the conclusion of this chapter, I provide some perspective based on the emergence of two contrasting types, the West African type and the Southeast Asian type, in the primitive accumulation process of the "periphery."

The human relations within the sub-Sahara-African rural community is disciplined by the principle of symbiosis, exemplified by the charity of the rich towards the poor.[145] This principle is for us at the origin of the economic stagnation that sub-Saharan Africa is experiencing. If foreign investments from Europe and other countries of the industrial "core" are stagnating in the sub-Saharan Africa, it is in particular because the principle of symbiosis slows down the processes of differentiation of peasant class and neutralizes social antagonisms within community.

On the contrary, in the agricultural villages of Southeast Asia, the differentiation of peasant class is more rapid, due to social relations based on individualism, with the consequence that farmers are much more in competition with each other, especially during the development of the Green Revolution. This process liberated a class of landless workers needed for industrialization. Thus, the combination of direct investment from foreign countries of the "core" and the breakdown of rural pre-capitalist elements, allowed industrialization to progress in Southeast Asia.

Rebuilding true symbiotic relationships in Southeast Asian villages does not mean reverting to symbiotic relationships like those found in sub-Saharan Africa. On the contrary, while this recovery takes as given premise the industrialization and productive forces brought about by the "core" throughout history, it takes place through the construction of a society free of the alienation, social inequality, and environmental destruction created by capitalist

145 Regarding "charity," see Chapter 4 of this book.

industrialization. Such a society is more likely to emerge in "peripheral" countries than in the "core," where capital relations are much stronger.[146]

146 Some may see the influence of the so-called frontier revolution theory by Otsuka Historiography as follows in the present author's statements in this sentence. "Although it is true that in the core regions of a given social formation new relations of production that characterize the next stage are created quickly, on the other hand, the foundations of the old relations of production remain so entrenched that the development of new modes of production will naturally be hindered or severely distorted. As a result, new modes of production naturally leave these central areas and migrate (or spread) to peripheral or adjacent regions where the formation of old production relations is relatively weak or almost absent. And there, on the contrary, new modes of production will achieve smooth and normal growth." (Otsuka 1960) However, rather than seeing similarities between the two (my statement and frontier revolution theory), I would rather hope that a comparison between the two will reveal the following problems. First, can we say that today, within the world system, whether locally or universally, a "new mode of production" is being created to replace the capitalist mode of production? Second, even if "new mode of production" is being created, can it be said to be growing and developing in the form of propagation from the "core" to the "periphery?" (How should we understand the historical significance of the economic development of socialist countries in East Asia?) Third, what role does the idea of symbiosis actually play in the above two processes? Or should it be done in the future?

Chapter 7 : Reexamination of Rosa Luxemburg's Schema of Reproduction: Or about the "Schema Incorporating Non-Capitalist Surroundings"

1. Theme

Rosa Luxemburg's (1913) "The Accumulation of Capital" argues that a capitalist society exists with the existence of a "non-capitalist surroundings" as an essential condition. Chapter 26, "The Reproduction of Capital and its Social Setting," develops this in particular detail. Chapter 26 cites the following three reasons why a capitalist society needs a "non-capitalist surroundings." One is the need for a "non-capital surroundings" to resolve the imbalance between sectors and the difficulty in realizing surplus-value that arise during the reproduction process. The second is the need for a "non-capitalist surroundings" as a source of raw materials and fuel. The third is the necessity of a "non-capitalist surroundings" as a source of labor supply.

By the way, when researchers have discussed "The Accumulation of Capital" in the past, they have mainly focused on the first of the three reasons mentioned above. It seems that the significance of Luxemburg's book has been denied while pointing out the insufficiency of the argumentation process. For example, Ichihara (1990) states:

"A. Pannekoek, who had a 'radical' position, also wrote a critical book review in the Bremer Bürger Zeitung (January 29, 1913). The 'centralists' of the Social Democratic Party of Germany (SPD) found a suitable material for counterattacking against the line of anti-imperialist struggle to which Rosa belonged, the 'radical faction.' That was a fallacy in the theory of reproduction on which her theory of imperialism was based. The 'centralists' tried to make the most of this opportunity."

As can be seen in this paragraph, the critique of the "reproduction theory" in "The Accumulation of Capital" was academic in nature, but at the same time it had a strong political flavor from the beginning. On the other hand, there are also arguments that have focused on reasons other than the first. For example, Bradby (1975) names the first reason Luxemburg's "strong thesis" and names the third reason, which has not received much attention, her "weak thesis." [147] However, this kind of discussion can be said to be an exception, and the debate surrounding the "The Accumulation of Capital" converged on the first reason while criticizing the insufficiency of its demonstrative process. Of course, there are also problems inherent in the description of "The Accumulation of Capital," which cannot be reduced to a purely political background. This is because "The Accumulation of Capital" itself emphasizes the first of the three reasons above, and does not seem to clearly explain the logical connection between the three reasons. This is why the first one is called a "strong thesis" by Bradby.[148]

Therefore, the first task of this chapter (in Sections 2 to 4) is to critically examine the three reasons stated in Chapter 26 of "The Accumulation of Capital." What I should keep in mind when doing so is, the considerations so far in this section lead me, to consider the second and third reasons as well, giving them their proper place. Through this work, I am able to present my

147 I learned about the existence of Bradby's point from Muroi (1995).

148 Yamazaki (2014b: final chapter), from the standpoint of seeing the fundamental contradiction of capitalist society in the commodification of labor-power, rather emphasizes the "weak thesis." On May 16, 2018, a joint review meeting for that book was held at the Faculty of Agriculture, Tokyo University of Agriculture and Technology. At that meeting, Professor Norio Tsuge of Tohoku University criticized my book's treatment of Luxemburg theory as follows. In my book, there is no unified understanding of Luxemburg theory. "The development of Yamazaki's book does not delve deeply into the problem of resolving the contradiction between production and consumption, but seems to focus on the problem of securing a labor force. Why is this?" The "first task" of this chapter, described below in the main text, is intended to respond to this comment.

own perspective on the question of whether a capitalist society exist with the existence of "non-capitalist surroundings" as an indispensable condition.

The second task is to consider the form of a schema of extended reproduction that incorporates the "non-capitalist surroundings" from the perspective obtained through the examination of the first task, and to tentatively propose this schema (in Section 5).

First, let us consider the first reason raised by Luxemburg.

2. Examination of Rosa Luxemburg's Schema of Extended Reproduction

Luxemburg develops her own argument while criticizing Marx's schema of extended reproduction. Therefore, at a slightly detour, let us first confirm the contents of Marx's (1885) schema of extended reproduction. Volume 2 of "Capital," Chapter 21, "Accumulation and Reproduction on an Extended Scale," Section 3, "Schematic Presentation of Accumulation" gives two examples of extended reproduction. Here, I will explain the changes from the first year to the third year in the second example, which is also discussed in Chapter 25 of "The Accumulation of Capital."[149] As pointed out by Marx, the annual progression in the reproduction schema merely indicates the passage of time, and does not necessarily correspond to the number of years in the calendar year. For example, the second year in the reproduction schema may be the tenth year in the calendar year. Thinking like this is necessary in order to treat the progress of accumulation while abstracting from the business cycle. First, the results of first year of production for the second illustration of Marx's schema of extended reproduction are as follows.

149 Luxemburg (1913) shows up to the fourth year, but here I will only give a simplified introduction up to the third year.

Second illustration of Marx's schema of extended reproduction

1st year

Department I 5,000c+1,000v+1,000m=7,000

Department II 1,430c+ 285v+ 285m=2,000

When preparing for the second year of production, the capital composition is assumed to remain unchanged at c:v = 5:1. Furthermore, half of the surplus-value in the first department goes to accumulation, and the other half goes to consumption by capitalists. Therefore, of the 1,000 surplus-value produced in the first department, only 500 goes to accumulation for the next year's production. Of this, 416.7, equivalent to 5/6, will be constant capital, and the remaining 83.3, equivalent to 1/6, will be variable capital. Therefore, the capital composition of the first department at the starting point in the second year is 5,416.7c + 1,083.3v.[150]

First, c in the second department has 1,500 derived from I (v+m/2) =IIc, which is the condition for equilibrium with the first department in the previous year. In addition, an amount corresponding to 83.3, which is the increase in v due to the accumulation of the first department, is added. Therefore, the constant capital of the second department in the second year is 1,583.3. The increase in v of the second department becomes 1/5 of the increase in c from the capital composition given as the precondition, i.e., 30.7. Therefore, the capital composition of the second department at the starting point in the second year is 1,583.3c + 315.7v.

The results of the production carried out after this preparation is as follows,

150 Because Marx rounds off fractions during calculations to obtain whole numbers, the numbers may differ slightly from those in this chapter, which calculates without rounding. The same can be said about the Luxemburg's schema of reproduction, which will be discussed later.

assuming that the rate of surplus-value is 100%, unchanged from the previous year.

2nd year

Department I　5,416.7c+1,083.3v+1,083.3m=7,583.3

Department Ⅱ　1,583.3c+　315.7v+　315.7m=2,214.7

Similarly, the third year will be as follows.

Department I　5,868.1c+1,173.6v+1,173.6m=8,215.3

Department Ⅱ　1,715.3c+　342.1v+　342.1m=2,399.4

Chapter 25, "Contradictions within the Diagram of Enlarged Reproduction," in "The Accumulation of Capital" examines the second illustration of Marx's schema of extended reproduction, and states the following points of view in order for criticizing Marx.

"There can be no doubt, therefore, that ①Marx wanted to demonstrate the process of accumulation in a society consisting exclusively of workers and capitalists, under the universal and exclusive domination of the capitalist mode of production. On this assumption, however, his diagram does not permit of any other interpretation than that of ②production for production's sake." (Underlining and ①② were added by the quoter.)

Here, ① does not require any special explanation. Regarding ②, specifically, it seems to refer to the following assumption. It is assumed that half of the surplus-value in the first department is accumulated, and the amount accumulated in the second department is determined to correspond to the amount accumulated in the first department so that there is just enough exchange between the two departments. Furthermore, Luxemburg argues that this assumption is possible because the following two points are not taken into account in the Marxian schema of extended reproduction.

(a) Marx's schema does not incorporate the increase of capital composition over time. In other words, the c:v ratio between the first and second departments remains unchanged at 5:1 no matter how much time passes.

(b) Also, the increase in the rate of surplus-value is not incorporated. That is, the surplus-value rate of the first and second departments is always 100%.

Pointing out these two deficiencies, Luxemburg creates a formula while considering the following three points.

(i) Capital composition increases progressively as capital increases. That is, c:v in the first year is 5:1, which is the same as Marx's constant ratio, but in the second year it is 6:1, in the third year it is 7:1, and in the fourth year it is 8:1. The ratio changes over time.

(ii) The rate of surplus-value increases in response to increases in labor productivity. Luxemburg does not explicitly state the figures in her text. However, when actually measured from the first department of the created schema, the rate of surplus-value she has adopted is 100% in the first year, the same as Marx's constant rate. However, the rate increases to 101.08% in the second year, 102.99% in the third year, and 105.48% in the fourth year, which is also increasing with time.

(iii) Just as Marx assumed, half of the surplus-value acquired in the first sector is capitalized.

The following is the schema of extended reproduction actually created by Luxemburg. Note that the values for the first year, which is the starting point, are the same as in the second illustration of Marx's schema of extended reproduction expression.

1st year

I　5,000c+1,000v+1,000m=7,000 (means of production)

Ⅱ　1,430c+　285v+　285m=2,000（means of consumption）　Total 9,000

2nd year

Ⅰ　5,428.6c+1,071.4v+1,083m=7,583（means of production）

Ⅱ　1,587.7c+　331.3v+ 316m=2,215（means of consumption）　Total 9,798

3rd year

Ⅰ　5,903c+1,139v+1,173m=8,215（means of production）

Ⅱ　1,726c+　331v+ 342m=2,399（means of consumption）　Total 10,614

4th year

Ⅰ　6,424c+1,205v+1,271m=8,900（means of production）

Ⅱ　1,879c+　350v+ 371m=2,600（means of consumption）　Total 11,500

Whereas Marx created numerical values in the reproduction schema so that the first and second departments were in equilibrium, Luxemburg here exclude the equilibrium relationship between the first and second departments. Under the condition that half of the surplus-value acquired in the first department is capitalized, the general formula for the equilibrium condition of extended reproduction takes the following form.

Iv+Im/2+⊿Iv=Ⅱc+⊿Ⅱc

From the first year to the second year in Luxemburg, the left side of this equation is 1,571.4, while the right side is 1,587.4, resulting in an imbalance of 16 between the two. This difference stems from the overaccumulation of the second department, set by Luxemburg herself. Next, from the second year to the third year, the left side of the formula for the equilibrium condition is 1,680.5, while the right side is 1,726, resulting in excess production in the second department of 46.5 (according to Luxemburg 45). Furthermore, from the third year to the fourth year, the left-hand side of the formula for the equilibrium condition is 1,791.5, while the right-hand side is 1,879, resulting in excess production in the second sector of 87.5 (according to Luxemburg 88).

Luxemburg claims that the reason why this imbalance occurs in the formula she created is because she takes into account the two conditions mentioned above that the second example of the Marxian schema of extended reproduction does not take into account. The two conditions were (i) an increase in the organic composition of capital and (ii) an increase in the rate of surplus-value. She states:

"On the basis of what is actually happening, namely a greater yearly increase of constant capital as against that of variable capital, as well as a growing rate of surplus-value, discrepancies must arise between the material composition of the social product and the composition of capital in terms of value."

However, even if all three of Luxemburg's conditions are taken into account, it is actually possible to create a balanced schema for extended reproduction. Therefore, below, I will show a balanced schema that I created as a prototype based on the three Luxemburg's conditions starting from the second year.

2nd year−part 2

The capital composition is 6:1, the rate of surplus-value is 101.08%, and the accumulation of the 1st department is half of the surplus-value of the previous year.

I　5,428.6c+1,071.4v+1,083m=　7,583 (means of production)

II　1,571.4c+　308.6v+　311.9m=2,192 (means of consumption) Total 9,775

From the 1st year to the 2nd year, ⊿v+Im/2+⊿Iv=IIc+⊿IIc=1,571.4

3rd year−part 2

The capital composition is 7:1, the rate of surplus-value is 102.99%, and the accumulation of the 1st department is half of the surplus-value of the previous year.

I 5,902.4c+1,139.1v+1,173.2m=8,214.7 (means of production)

II 1,680.6c+ 324.2v+ 333.9m=2,339 (means of consumption) Total 10,553.7

From the 2nd year to the 3rd year, Iv+Im/2+⊿Iv=IIc+⊿IIc=1,680.6

4th year−part 2

The capital composition is 8:1, the rate of surplus-value is 105.48%, and the accumulation of the 1st department is half of the surplus-value of the previous year.

I 6,423.8c+1,204.3v+ 1,270.3m=8,898.4 (means of production)

II 1,790.9c+ 338v + 356.5m=2,485 (means of consumption) Total 11,383.4

From the 3rd year to the 4th year, ⊿v+Im/2+⊿Iv=IIc+⊿IIc=1,790.9

However, in the Luxemburg's illustration, the capital composition mentioned above is adopted as the capital composition of the first department, and a different capital composition is actually adopted for the second department. Luxemburg herself does not indicate this, so measured based on the schema she created, the ratio is 4.79:1 in the 2nd year, 5.21:1 in the 3rd year, and 5.37:1 in the 4th year. As you can see, all of these numbers are around 5:1, hence Luxemburg may be trying to adopt 5:1. However, here we will recalculate the balanced schema from the 2nd year onwards based on these actual measurements as follows.

2nd year−part 3

The capital composition of the 1st department is 6:1, the capital composition of the 2nd department is 4.79:1, the rate of surplus-value is 101.08%, and the accumulation of the 1st department is half of the surplus-value of the previous year.

Ⅰ $5{,}428.6c+1{,}071.4v+1{,}083m=7{,}583$ (means of production)

Ⅱ $1{,}571.4c+314.5v+317.9m=2{,}204$ (means of consumption) Total 9,787

From the 1st year to the 2nd year, $\Delta v+Im/2+\Delta Iv=IIc+\Delta IIc=1{,}571.4$

3rd year−part 3

The capital composition of the 1st department is 7:1, the capital composition of the 2nd department is 5.21:1, the rate of surplus-value is 102.99%, and the accumulation of the 1st department is half of the surplus-value of the previous year.

Ⅰ $5{,}902.4c+1{,}139.1v+1{,}173.2m=8{,}214.7$ (means of production)

Ⅱ $1{,}680.6c+335.5v+345.5m=2{,}362$ (means of consumption) Total 10,576.7

From the 2nd year to the 3rd year, $Iv+Im/2+\Delta Iv=IIc+\Delta IIc=1{,}680.6$

4th year−part 3

The capital composition of the 1st department is 8:1, the capital composition of the 2nd department is 5.37:1, the rate of surplus-value is 105.48%, and the accumulation of the 1st department is half of the surplus-value of the previous year.

Ⅰ $6{,}423.8c+1{,}204.3v+1{,}270.3m=8{,}898.4$ (means of production)

Ⅱ $1{,}790.9c+356v+375.5m=2{,}522$ (means of consumption) Total 11,420.4

From the 3rd year to the 4th year, $\Delta v+Im/2+\Delta Iv=IIc+\Delta IIc=1{,}790.9$

Comparing the results of these two series of calculations by the present author with the numerical example presented in “The Accumulation of Capital,” the numerical values for the first department are almost the same among the three series. However, the figures for the second department vary widely. Now, if we extract only the figures for the second year of the second department and juxtapose them, we get the following.

2nd year

Ⅱ 1,587.7c+331.3v+316m=2,215 Total 9,798 "The Accumulation of Capital"

Ⅱ 1,571.4c+308.6v+311.9m=2,192 Total 9,775 Present author (part 2)

Ⅱ 1,571.4c+314.5v+317.9m=2,204 Total 9,787 Present author (part3)

The difference between (part 2) and (part 3) is the capital composition of the second department, but it is clear from the following point that this difference does not affect the equilibrium conditions. Looking at the above general formula regarding the equilibrium conditions for extended reproduction, the capital composition of the second department is not involved. This comparison of Luxemburg's and the present author's illustrations shows that the sectoral imbalance in Luxemburg's schema of extended reproduction does not result from her taking into account the two factors she herself mentions; (i) increase in the capital composition and (ii) increase in the rate of surplus-value. Rather, this imbalance arises solely from arbitrary numerical manipulations that make the accumulation in the second department too much so that equilibrium with the first department does not hold.[151] Marx's schema of reproduction has positive significance in that it attempted to demonstrate the viability of capitalist society while criticizing Adam Smith's v+m dogma by showing the quantitative proportional relationship between two departments. If this is the case, then arbitrarily setting the values in the schema and claiming that unbalanced relationships between departments cannot be said to

151 Similar points were made by Yamada (1951) and Ichihara (1990).

be a correct understanding of the issues raised by Marx's schema.[152]

However, despite this, "The Accumulation of Capital" proceeds as follows at the beginning of Chapter 26.

"Marx's diagram of enlarged reproduction cannot explain the actual and historical process of accumulation. And why? Because of the very premises of the diagram. The diagram sets out to describe the accumulative process on the assumption that the capitalists and workers are the sole agents of capitalist consumption. We have seen that Marx consistently and deliberately assumes the universal and exclusive domination of the capitalist mode of production as a theoretical premise of his analysis in all three volumes of 'Capital.' Under these conditions, there can admittedly be no other classes of society than capitalists and workers; as the diagram has it,"

---

152 Bauer (1913) created a schema that "removed arbitrariness" from the Marx schema for the purpose of criticizing Luxemburg. That is, schema is created under the following five conditions. ① The annual increase rate of population (variable capital) is 5%. ② The growth rate of constant capital is 10% per year. ③ The rate of surplus-value is 100%. ④ Capital composition rises. ⑤ The rate of capital accumulation is the same between the two departments. Furthermore, Bauer solves the imbalance between departments seen in his schema by transferring capital between departments: Capitalists in the 2nd department invest in the 1st department. Luxemburg's (1921) anti-criticism against Bauer is as follows. (a) Capital movement between departments undermines the significance of the Marxian division into two departments. (b) It is unrealistic to think that capital accumulation is determined by the natural rate of population increase. Furthermore, Eckstein (1913) states that Luxemburg assumes the same accumulation rate between the two departments. The present author actually measured the accumulation rate that Luxemburg actually uses in the 2nd department, and found that it was 72% in the 1st → 2nd year, 44% in the 2nd → 3rd year, and 50% in the 3rd → 4th year. These figures differ from the permanent rate of 50% for the 1st department, except for the figure of the 3rd → 4th year. Furthermore, Luxemburg (1913) states that in the capital accumulation from the 1st to the 2nd year of the 2nd department in her own schema, she adopted the same 184 capitalization as in the 2nd illustration of the Marx schema of extended reproduction. On the other hand, she seems to have created the schema with an eye toward keeping the ratio between the 1st and 2nd departments constant at 3.4 to 3.5.

"They (workers and capitalists—citer) can always only realize the variable capital, that part of the constant capital which will be used up, and the part of the surplus-value which will be consumed, but in this way they merely ensure that production can be renewed on its previous scale. The workers and capitalists themselves cannot possibly realize that part of the surplus-value - which is to be capitalized. Therefore, the realization of the surplus-value for the purposes of accumulation is an impossible task for a society which consists solely of workers and capitalists."

"The decisive fact is that the surplus-value cannot be realized by sale either to workers or to capitalists, but only if it is sold to such social organizations or strata whose own mode of production is not capitalistic."

The following criticism has been made of Luxemburg's reasoning. "In one fell swoop, Rosa leaps to the false position that the sale of the accumulated portion of surplus-value cannot take place without a 'non-capitalist surroundings'." (Tomizuka 1990) Whether or not focusing on the "non-capitalist surroundings" is a "leap," it cannot be denied that there are difficulties with Luxemburg's reasoning. Above all, the reasons for the need for a "non-capitalist surroundings" are different between Chapters 25 and 26, and in that sense the arguments are inconsistent. As we saw earlier, what Chapter 25 discusses is the intersectoral imbalance between the 1st and 2nd departments due to overaccumulation and overproduction in the 2nd department. On the other hand, according to the above quotation in Chapter 26, in the extended reproduction of a capitalist society consisting only of capitalists and workers, surplus-value (the portion of which should be capitalized that exceeds the consumption portion of capitalists) is impossible to realize. These cited sentences state that even when equilibrium is established between departments, it is generally difficult to realize a portion of surplus-value. This inconsistency between Luxemburg's Chapters 25 and 26 may actually indicate two separate reasons why a capitalist society needs a "non-capitalist surroundings," although she does not explicitly state this. However, even if this were the case, each of these two reasons would have to be denied; (α) imbalance between departments and (β) difficulty in selling a portion of surplus-value. This is because, first of all, the factor that led to the "(α) imbalance between departments" is Luxemburg's arbitrary setting of numerical values for the 2nd department that are suitable for drawing her own conclusions, as we have already seen. Regarding "(β) difficulty in selling a portion of surplus-value," Marx's schema of extended reproduction already shows that surplus-value can be sold in just the right amount during extended reproduction. On this point too, Luxemburg's point is not correct.

3. Other Reasons Why Capitalist Production Requires a "Non-Capitalist Surroundings"

By the way, Chapter 26 of "The Accumulation of Capital" discusses three reasons why capitalist production requires a "non-capitalist surroundings," as already mentioned. However, this is only if the above two reasons are allowed to be lumped together as a sales issue. Of these three, we looked at the first sales issue in detail in the previous section, so let us consider the remaining two here.

First, let us talk about the second issue; the need for "non- capitalist surroundings" as a source of raw materials and fuel. On this point, Luxemburg (1913) states as follows:

"Moreover, capitalist production, by its very nature, cannot be restricted to such means of production as are produced by capitalist methods. Cheap elements of constant capital are essential to the individual capitalist who strives to increase his rate of profit. In addition, the very condition of continuous improvements in labor productivity as the most important method of increasing the rate of surplus-value, is unrestricted utilization of all substances and facilities afforded by nature and soil. To tolerate any restriction in this respect would be contrary to the very essence of capital, its whole mode of existence. After many centuries of development, the capitalist mode of production still constitutes only a fragment of total world production. Even in the small Continent of Europe, where it now chiefly prevails, it has not yet succeeded in dominating entire branches of production, such as peasant agriculture and the independent handicrafts; the same holds true, further, for large parts of North America and for a number of regions in the other continents."

I will explain this point later in conjunction with the third point, but for now, let us look at the third point, which is related to labor procurement. Luxemburg states:

"Labor for this army is recruited from social reservoirs outside the dominion of capital – it is drawn into the wage proletariat only if need arises. Only the existence of non-capitalist groups and countries can guarantee such a supply of additional labor-power for capitalist production."

Earlier, I wrote about "three reasons" why capitalist production requires "non-capitalist surroundings." However, Luxemburg's writing style does not actually give equal weight to all three factors. Regarding the first sales problem, as we saw earlier, the necessity of "non-capital surroundings" is derived from it. Regarding the third issue of additional labor-power, Luxemburg says that "only the existence of non-capitalist groups and countries can guarantee such a supply of additional labor-power for capitalist production." Therefore, it can be said that she derives the necessity of "non-capitalist surroundings" from this. However, unlike these two reasons, regarding the second problem of raw material and fuel procurement, it is written that "capitalist production, by its very nature, cannot be restricted to such means of production as are produced by capitalist methods." In other words, the necessity of "non-capital surroundings" is not derived from this. Rather, the argument is that capitalist production utilizes the primary products provided by given "non-capitalist surroundings" if the primary products are cheap and contribute to an increase in surplus-value. Another argument is that it is impossible to leave such products unused. Indeed, Chapter 26 of "The Accumulation of Capital" deals with the following three main channels through which capitalist production comes into contact with the "non-capitalist surroundings" through the market; a sales channel for capitalist products, a source of raw materials and fuel, and a source of labor supply. However, from the perspective of the necessity of the "non-capitalist surroundings," it can be seen that the focus is narrowed down to two things; a sales channel for capitalist products and a source of labor supply.

Of these two, Luxemburg deduces the necessity of the "non-capitalist surroundings" as a "sales channel for capitalist products" using a certain procedure from the reproduction schema. However, as we saw earlier, there were some difficulties. However, regarding the necessity of the "non-capitalist surroundings" as a "source of labor supply," the proposition cited earlier appears rather abruptly. The argumentative process for deriving this proposition cannot be found in "The Accumulation of Capital." This is probably why Bradby called the former Luxemburg's "strong thesis" and the latter its "weak thesis."

From the examination of "The Accumulation of Capita" so far, all three of Luxemburg's arguments for the necessity of the "non-capitalist surroundings" for capitalist production lack persuasiveness. Is it, thus, completely meaningless to consider the necessity of the "non-capitalist surroundings" for capitalist production? Here, we will consider this issue further by following Luxemburg's example and considering the schema of extended reproduction.

4. Reconsidering Marx's Schema of Extended Reproduction

The fact to note here is that Marx's and Luxemburg's schemas of extended reproduction assume an increase in variable capital over time.[153] And this is probably something that anyone would notice if they looked at both schemas. However, in previous research, I think that what this means has not been sufficiently considered. Therefore, here I would like to consider what this fact means.

First, to confirm the facts, let's take another look at Marx's (1885) second example of the schema of extended reproduction, and examine how the size of variable capital changes there. It progresses as follows.

Increase in variable capital in the 2nd example of Marx's schema of extended reproduction

1st year (I) 1,000v+ (II) 285v=1,285v

2nd year (I) 1,083v+ (II) 316v=1,399v

Increase rate of variable capital is 8.9%.

3rd year (I) 1,173v+ (II) 342v=1,515v

Increase rate of variable capital is 8.3%.

4th year (I) 1,271v+ (II) 371v=1,642v

Increase rate of variable capital is 8.4%.

In this case, the annual increase rate of variable capital is 8-9%. Next, let us extract the part of variable capital of the Luxemburg's schema, which is

153 Lenin (1893) gave a numerical example of extended reproduction that does not assume an increase in variable capital, but then dismissed this example as "based on an unlikely assumption and therefore incorrect."

another schema of extended reproduction that this chapter deals with.

Increase in variable capital in Luxemburg's schema of extended reproduction

1st year	(I) 1,000v+ (II) 285v=1,285v	
2nd year	(I) 1,071v+ (II) 331v=1,403v	
		Increase rate of variable capital is 9.2%.
3rd year	(I) 1,139v+ (II) 331v=1,470v	
		Increase rate of variable capital is 4.8%.
4th year	(I) 1,205v+ (II) 350v=1,555v	
		Increase rate of variable capital is 5.8%.

Like Marx, Luxemburg also assumes that variable capital will increase year by year, but the rate of increase varies from about 5% to about 9%. In any case, this increase in variable capital must ultimately mean an increase in the number of workers. The reason I say "ultimately" here is because it is also possible to consider the increase in wages associated with longer working hours and increased labor intensity. In this case, it is possible to increase the labor supply without increasing the number of workers. However, there are limits to these things, conditioned by the natural and physical constraints of the workers. On the other hand, the natural increase in the labor force itself can also be considered. Thus Marx did not ignore this element.[154] In fact,

154 The following statement in Chapter 22 of Volume 1 of "Capital," "Conversion of Surplus-Value into Capital," emphasizes the significance of the increase in the number of workers through reproduction in capital accumulation. "Now in order to allow of these elements (part of surplus-value—citer) actually functioning as capital, the capitalist class requires additional labor. If the exploitation of the laborers already employed do not increase, either extensively or intensively, then additional labor-power must be found. For this the mechanism of capitalist production provides beforehand, by converting the working-class into a class dependent on wages, a class whose ordinary wages suffice, not only for its maintenance, but for its increase. It is only necessary for capitalist to incorporate

capitalist society improved medical care and hygiene, leading to rapid population growth. However, an important implication of the theory of capital accumulation is that capital accumulation can be achieved by overcoming the constraint of natural increase in labor-power.

Therefore, let us look for theories of labor supply other than those mentioned above in "Capital." This leads us to the theory of relative surplus-population, which is developed in Chapter 23 of Volume 1, and the theory of primitive accumulation, which is developed in Chapter 24. As is well known, relative surplus-population is an argument that the increase of capital composition and recession create surplus-population as a source of labor supply. Furthermore, the theory of primitive accumulation argues that direct producers such as peasants transform into workers by losing land and other means of production, and this process creates a supply of labor. There are these two theories, but the relationship between them is not necessarily obvious. In addition, the relative surplus-population, in which the increase of the capital composition brings about the supply of labor, is somewhat difficult to understand. This is because, no matter how increased the capital composition becomes, as long as capital accumulation progresses, the demand for labor in capitalist sector will only increase, and labor may not be supplied from there. Let us demonstrate this using a simple numerical example. Now, assume that the average composition of total social capital is c:v. In that case, even if this ratio changes from 2:1 to 4:1, if the amount of total social capital increases by more than 5/3 times over the same period, no surplus-population will be created. On the contrary, the demand for labor will increase. In this example, surplus-population is formed in a special case where the increase in

---

this additional labor-power, annually supplied by the working-class in the shape of laborers of all ages, with the surplus means of production comprised in the annual produce, and the conversion of surplus-value into capital is complete."

total social capital is less than 5/3 times, so the argument that an increase in the capital composition creates surplus-population does not hold in general. This argument only holds true under special conditions.[155]

Therefore, on the one hand, the question is how to think about the relationship between the two labor supply theories mentioned above. On the other hand, the question is what are the conditions under which labor supply can be achieved through the formation of a relative surplus-population. I would like to introduce Miyazawa (2018) who has developed an interesting discussion on these points. He divides the factors of labor supply into three with different time axes; (1) those associated with business cycles, (2) those associated with the rise of capital composition, and (3) labor supply from agriculture. Of these, it is said that the condition for a cumulative increase in relative surplus-population is that the labor supply is active due to factor (3). In other words, he argues that the relative surplus population ((1) and (2)) and the primitive accumulation ((3)) run concurrently, although with a time limit of "Marx's era." He argues that the existence of such a synchronic structure is a condition for the relative surplus-population to be a labor supply theory.

I affirm this three-dimensional understanding of the causes of formation of relative surplus-population, but at the same time believe the historical transition of labor supply according to the stages of development of capitalist society; (3) → (1) and (2). In other words, what I would like to argue is the theory of historical changes in the labor supply mechanism from the period of primitive accumulation to the period of formation of relative surplus-

155 Okishio (1980) mathematically demonstrates the proposition that when "the capital composition becomes sufficiently increased," "a cumulative surplus-population is formed." However, in reality, "organic composition does not show a sufficient tendency to increase." Arai (1985) has criticized Okishio's argument.

population.[156] Furthermore, this change is also suggested by the texts of "Capital." [157]

By the way, based on the above reasoning, the following thought is not far from Marx's true meaning. That is, in "extended reproduction = capital accumulation," the dissolution of the "non-capitalist surroundings" or the process of primitive accumulation is an essential element as a source of labor-power. For Miyazawa, the supply of labor-power from the "non-capitalist surroundings" to the capitalist sector was a prerequisite for establishing a relative surplus- population. I believe the historical transition of the labor supply mechanism in the capitalist sector from the period of primitive accumulation to the period of relative surplus-population. If so, the "non-capitalist surroundings" was a major source of labor during the period of primitive accumulation. In addition, I believe that during the period of relative surplus-population, the situation will become more complex as the effects of the

156 For details, see Chapter 1 of this book.

157 From the following passage in Chapter 24 of Volume 1 of "Capital," "The So-called Primitive Accumulation," we can read the theory of historical development from "the period of primitive accumulation to the period of formation of relative surplus-population." "The organization of the capitalist process of production, once fully developed, breaks down all resistance. The constant generation of a relative surplus-population keeps the law of supply and demand of labor, and therefore keeps wages, in a rut that corresponds with the wants of capital. The dull compulsion of economic relations completes the subjection of the laborer to the capitalist. Direct force, outside economic conditions, is of course still used, but only exceptionally. In the ordinary run of things, the laborer can be left to the 'natural laws of production,' i.e., to his dependence on capital, a dependence springing from, and guaranteed in perpetuity by, the conditions of production themselves. It is otherwise during the historic genesis of capitalist production. The bourgeoisie, at its rise, wants and uses the power of the state to 'regulate' wages, i.e., to force them within the limits suitable for surplus-value making, to length the working-day and to keep the laborer himself in the normal degree of dependence. This is an essential element of the so-called primitive accumulation." On the other hand, during the period of primitive accumulation, agriculture and rural areas were the main source of labor, which was a factor in low wages.

rise of the capital composition come into play, but the supply of labor from the "non-capitalist surroundings" will continue. Furthermore, let us look back at one classic argument. In the debate between Tolokonski et al. and Oppenheimer over the sources of labor supply associated with capital accumulation, the last thing Tolokonski et al. brought up was the decomposition of the "non-capitalist surroundings." To begin with, Marx's schema of extended reproduction assumes an increase in variable capital and the worker population.[158]

Based on the above, I believe that the schema of extended reproduction includes the absolute increase in the working population. Furthermore, unless we assume that this increase is all due to population growth due to reproduction (although population growth due to reproduction cannot be ignored), we cannot help but think that the process of separation between direct producers, mainly farmers, and the means of production (mainly land) is progressing along with capital accumulation. The schema of extended reproduction can be theorized by assuming the situation of a capitalist society that coexists with primitive accumulation. Once we accept this, the following theoretical questions arise.

If the schema of extended reproduction is a theoretical model of a capitalist society that coexists with primitive accumulation, does it indicate the state of a transitional capitalist society before it becomes "purified?" Alternatively, there may be the following theoretical position. The schema of extended reproduction is a theoretical model of a capitalist society that coexists with

158 Marx (1862-1863) says: "The absolute increase in population (even though it is decreasing relative to the capital employed) is a condition for accumulation to be a continuous process. Population growth appears as the basis of an ever-evolving process of accumulation." Furthermore, Luxemburg (1921), quoting these texts, states: "Without an increasing population of workers, continuous expansion of production cannot take place. Moreover, even the simplest workers understand this."

primitive accumulation, but isn't that what a capitalist society is like in the first place? This is the theoretical position that a capitalist society can only exist if it coexists with primitive accumulation. From the description of "The Accumulation of Capital" mentioned above, Luxemburg appears to be adopting the latter position. Or, it seems that the position of Tolokonski who defended Marx in the classic debate with Oppenheimer, a critic of Marx, was similar. I believe that, when viewed on a global scale, capitalist society is based on the coexistence with primitive accumulation, but as a single country, capitalist society is possible to reach the ultimate limit of "purification." Having reached the limit of such "purification," a single-country capitalist society begins to expand overseas in search of the "non-capitalist surroundings" as a source of labor for the capitalist sector. In other words, such a society develops with the introduction of immigrants and the export of capital.

Based on the above considerations, in what direction should the discussion proceed from here? In that case, one meaningful direction would be to consider the form of a schema of extended reproduction that incorporates the "non-capitalist surroundings." "The Accumulation of Capital" places great emphasis on the necessity of the "non-capitalist surroundings" for capitalist production. Furthermore, Luxemburg's book makes heavy use of the schema of extended reproduction as a tool to demonstrate this theoretical proposition. However, for some reason, no attempt has been made to create a schema of extended reproduction that incorporates the "non-capitalist surroundings" located at the intersection of the two. However, the necessity of incorporating the "non-capitalist surroundings" into the reproduction schema should not be derived from the sales problem and "the perspective of the purchased goods market as a sales channel for capitalist products," which is emphasized by Luxemburg. This is because this direction has already been denied in Section 2 of this chapter. Rather, the necessity of incorporating the "non-capitalist surroundings" into the reproduction schema should be derived from the problem of labor supply sources for capital, that is, from a "labor market perspective." [159] Luxemburg sees the basic contradiction of capitalist society as the contradiction between production and consumption that stems from the anarchy of this society,[160] but I see the basic contradiction of capitalist society as the contradiction (impossibility) of commodifying labor-power. This difference in the basic viewpoints of the two may also be reflected in the process of deriving necessity of incorporating the "non-capitalist surroundings." Therefore, what is the schema of extended reproduction that incorporates the "non-capitalist surroundings?"

5. Schema of Extended Reproduction Incorporating "Non-capitalist Surroundings"

What do we need to consider when designing a schema of extended

159 Uno (1932) limits the problems of reproduction schema as follows. "The fact that a general overview of capitalist production can be given by a few simple mathematical formulas does not mean that every moment of capitalist production can be expressed numerically. It is simply an expression of the basic relationships that are generally common to social reproduction as a mathematical formula according to a particular form of capitalism." Furthermore, Hidaka (1983) places accumulation theory after theory of reproduction schema, stating as follows: "Although the reproduction schema can express physical relations, it cannot express the reproduction of labor-power, and must only assume this. Therein lies the limit of the reproduction schema. / Let us consider the accumulation of capital as a more concrete example of the reproduction process." The answer suggested by Hidaka's theory of accumulation is the formation of surplus-population due to the rise of capital composition. My position differs from Hidaka's "Economic Principles" in that I attempt to express the absolute increase in the working population in the reproduction schema.

160 "From the point of view of scientific socialism, the historical necessity of socialist revolution manifests itself above all in the growing anarchy that drives capitalism into a dead end with no way out." (Luxemburg 1899)

reproduction that incorporates the "non-capitalist surroundings?" Here, the "non-capitalist surroundings" are represented by the peasant agricultural sector. In this case, we should refer to Marx's (1894) frequently quoted statement regarding peasant owning a parcel.

"For the peasant owning a parcel, the limit of exploitation is not set by the average profit of capital, in so far as he is a small capitalist; nor, on the other hand, by the necessity of rent, in so far as he is a landowner. The absolute limit for him as a small capitalist is no more than the wages he pays to himself, after deducting his actual costs. So long as the price of the product covers these wages, he will cultivate his land, and often at wages down to a physical minimum."

"This is one of the reasons why grain prices are lower in countries with predominant small peasant land proprietorship than in countries with a capitalist mode of production. One portion of the surplus-labor of the peasants, who work under the least favorable conditions, is bestowed gratis upon society and does not at all enter into the regulation of price of production or into the creation of value in general. This lower price is consequently a result of the producers' poverty and by no means of their labor productivity."

Peasant owning a parcel is peasant who farm with family labor on his own land. In a free competitive society where, capitalist production is dominant and agricultural production is mostly carried out by capitalist farms, according to Marx's theory of rent, the price of agricultural products in the market will be formed as follows: This price is the cost of the production of marginal land necessary to satisfy social demand, plus average profit and absolute rent. Furthermore, differential rents are added to the prices of agricultural products produced under more advantageous agricultural conditions. However, in places where agriculture is generally carried out by peasants, these economic categories have not yet been socially established. Peasants can continue agricultural production if they can secure enough wages from their own labor to sustain their livelihood through the sale of agricultural products. Therefore, the prices of agricultural products supplied by peasants can be lower than those of capitalist agriculture. However, this is not due to the high productivity of agricultural labor, but due to the distressed sales by peasants.

However, when incorporating the "non-capitalist surroundings" into the reproduction schema, the conditions that need to be taken into account are not only the fact that peasants tend to be satisfied with only receiving the wage, as just mentioned. In the remainder of this chapter, I do not assume a natural increase in the working population, and assume that the supply of labor to the capitalist sector is achieved exclusively through the disintegration of the non-capitalist agricultural sector into capitalists and workers.[161] Hence the total value of products produced in the non-capitalist agricultural sector will decline over time.[162] This is a characteristic of the non-capitalist agricultural sector, which is different from the schema of extended reproduction since Marx, in which the total value of products increases in both the 1st and 2nd departments in the capitalist sector. The third condition to be considered is that part of the non-capital agricultural sector is a production department of

---

161 In the "demographic transition," which depicts long-term population dynamics, high births and high deaths in the first period lead to high births and low deaths in the second period, and then lead to low births and low deaths in the third period. (Watanabe 1996) In this chapter, natural population growth is abstracted away to avoid a long schema.

162 In this chapter, for the sake of simplicity, I ignore changes in labor wages over time. In addition to what was mentioned in Note 157 (decreased wages due to relative surplus-population, low wages in agriculture and rural areas), what effect does the development of capitalist society have on labor wages? First, labor wages have a potential tendency to decline due to cheaper means of living due to technological innovation. On the other hand, since the range of means of livelihood depends on the "cultural stage of a country," wages have the potential to rise along with social development. Actual wage movements are determined by the mutual offsetting influences of these factors.

production means, and the remaining part is a production department of consumption means. It is true that most agricultural products are food, which is a means of consumption, but they also include industrial raw materials such as cotton, cocoons, and raw materials for processed foods, so agriculture also produces means of production.

Taking these conditions into account, I created the schema of extended reproduction that incorporates the "non-capitalist surroundings" in **Table 7-1.** The more specific rules used when creating this schema are shown below.

① In this schema, in order to make it possible to "typically and purely represent the forms which capital takes and discards" (Yamada 1951) in the distribution process, I assume exchange according to value, following Marx's schema. That is, value revolutions, business cycles, and production prices are ignored. However, agricultural products are assumed to be exchanged on a c+v basis for the reasons mentioned above. Also, following Marx's schema, surplus-value is expressed in its original form rather than in its transformation form (profit, interest, commercial gain, rent, etc.).

② I and II indicate the 1st department and 2nd department, and a and b indicate the capitalist sector and the peasant sector. Therefore, the combination of Roman numerals and alphabets represents the combination of departments and modes of production. That is, Ia is the capitalist 1st department, Ib is the peasant 1st department, IIa is the capitalist 2nd department, and IIb is the peasant 2nd department. The use of Roman numerals to represent departmental differences is a tradition dating back to Marx, but the use of alphabetic characters to represent differences of modes of production is unique to this chapter.

③ c is constant capital, v is variable capital, and m is surplus-value. Any unit of currency may be used. These are also traditions that date back to Marx.

④ The rate of surplus-value is always 100%. In other words, the increase in the rate of surplus-value over time that Luxemburg claims is not adopted here. The 100% rate of surplus-value is the one adopted by Marx and Lenin.

⑤ The rate of capital accumulation for Ia is always 70%. In other words, 70% of the surplus-value produced by Ia becomes the source of capital accumulation for the next year. This figure is higher than the 50% rate of capital accumulation in the 1st department that has traditionally been used since Marx. By changing this value, the rate of capital accumulation in society becomes faster than if no changes were made. As we will see later, this chapter shows the possibility of using the reproduction schema as a tool for analyzing the "historical changes = long-term dynamic changes" of capitalist society. In order to express these long-term changes in a form that is as compressed as possible, it is convenient to show the rapid accumulation of capital and the accompanying social changes in a short period of time. The reason why the rate of capital accumulation for Ia is raised to 70% here, instead of Marx's 50%, is to compress and express changes over time in the reproduction schema. The reproduction schema in this chapter is shown up to the 12th year. However, as the rate of capital accumulation in Ia declines, the speed of capital accumulation in society also declines, and the number of years required to extract the points discussed later will further increase.

⑥ The capital composition, c:v, in the 1st year , is 4:1 in the Ia and 2:1 in the IIa. The values of these ratios are the same as in the first example of Marx's schema of extended reproduction. In this chapter, the capital composition of the accumulated portion when capital is accumulated also initially maintains these ratios. However, in the 4th year, two cases are shown; a case in which the capital composition remains unchanged (part 1), and a case in which the composition of the portion of newly accumulated capital becomes more raised (part 2). Furthermore, from the 5th to the 10th year, the capital composition of the portion of newly accumulated capital will gradually become more raised as the years pass. Limiting the rising of capital composition to the newly accumulated portion of capital is the same as Luxemburg's and Lenin's methods. In addition, the reason why this chapter introduces the element of rising of the capital composition into the 4th year is because if it were not done, there would be a labor shortage. Conversely, up to the 4th year, I do not assume that the capital composition will become more raised. This is because there is an abundant supply of labor from peasants to the capitalist sector, which is thought to put a brake on the adoption of labor-saving technologies. It is assumed that the capital composition will be stagnant from the 11th year onward, and this operation is intended to express the stagnation of technological development that accompanies the expansion of fixed capital.

⑦ The products of the peasant sector in Ib and IIb have no surplus-value due to the first condition shown above. Therefore, the value composition of the products of the peasant sector is only c + v, and the m part, which exists in the capitalist sector, is missing. Moreover, the c:v of the products of the peasant sector is always assumed to be 1:1. In other words, it is assumed that the magnitude of c relative to v is always the same and does not increase over time.

⑧ The method of calculating accumulation in the capitalist sector in this chapter is similar to the calculation method used by Marx and Lenin when creating the schema of extended reproduction. However, some modifications have been made to take into account the existence of the peasant sector. Therefore, here I will express the progression of Marx's and Lenin's schema of extended reproduction in a general form. First, the calculation method for the 1st example of Marx's schema of extended reproduction is shown in a general form as follows.

Xth year

I　Ac+Bv+Cm=D

II　Ec+Fv+Gm=H

(X+1) th year

I　(A+0.4C) c+ (B+0.1C) v+ (B+0.1C) m=I

II　(B+0.6C) c+ (0.5B+0.3C-0.5E+F) v+ (0.5B+0.3C-0.5E+F) m=J

Using these formulas, we can accurately reproduce the progression of the 1st example of the Marxian schema of extended reproduction. In this 1st example, it is assumed that the rate of capital accumulation is 50%, the capital composition (c:v) of the 1st department is 4:1, and the capital composition of the 2nd department is 2:1. The coefficient values expressed using A, B, C, E, F, and G in the formulas also reflect these values of the rate of capital accumulation and capital composition. In the 2nd example of Marx's schema of extended reproduction and in the Lenin's schema of extended reproduction, the rate of capital accumulation is 50%, which is the same as in the 1st example, but the capital compositions change. Considering the changes in capital compositions, the coefficients in the formulas will also change. In order to respond to these changes in capital composition, it is necessary to express the formula in a more general form, expressing the coefficients in the formula as a function of the rate of capital accumulation and capital composition. Therefore, I show such a general formula below.

(X+1) th year (part 2)

1st department: Production of production-goods

$(A+rs_1C)$ c + $[B+r(1-s_1)C]$ v + $[B+r(1-s_1)C]$ m=I

2nd department: Production of consumer goods

$[B + (1\text{-}rs_1) C] c + [F + \{B + (1\text{-}rs_1) C\text{-}E\} \ (1/s_2\text{-}1)] v + [F + \{B + (1\text{-}rs_1) C\text{-}E\} \ (1/s_2\text{-}1)] \ m = J$

Note that the rate of capital accumulation is r, the capital composition (c:v) of 1st department is s_1: $(1\text{-} s_1)$, and the capital composition of 2nd department is s_2: $(1\text{-} s_2)$.

Using these general formulas, we can also reproduce Lenin's schema of extended reproduction and the 2nd example of Marx's schema of extended reproduction. However, as has already been pointed out by many, Lenin's scheme of extended reproduction contains a calculation error.

⑨ The schema of extended reproduction that incorporates the "non-capitalist surroundings" that this chapter is proposing takes the following form. Note that this schema also includes equations that show the value composition of the peasant sector. I will show later how the value composition of the peasant sector will change.

Xth year

Ia Ac+Bv+Cm=D

Ib Kc+Lv=M

IIa Ec+Fv+Gm=H

IIb Nc+Pv=Q

What forms will Ia, Ib, IIa, and IIb take in the next year, that is, in the (X+1) th year? Let us take a look at them one by one below.

⑩ First, Ia: (X+1) th year of the capitalist production department of production-goods is the same as the general formula for the 1st department, which was already derived from the Marxian schema of extended reproduction without incorporating non-capitalist factors. That is, it takes the following form.

Note that the meanings of the symbols r, s_1, and s_2 used below are the same as described above.

Department Ia: (X+1) th year of the department of capitalist production of production-goods

$(A+rs_1C)\,c + [B+r(1\text{-}s_1)C]\,v + [B+r(1\text{-}s_1)C]\,m = R$

⑪ Next, we look at Ib and IIb, that is, the department of peasant production of production-goods and the department of peasant production of consumer goods. In the two departments, as mentioned earlier, decomposition of peasant class occurs as time progresses. Peasant decomposition is the division of peasants into workers and capitalists. Therefore, with the decomposition of the peasant class, the number of peasants and the value of their products will decrease, and some agricultural products will instead be produced in a capitalist manner. If we try to understand these things as changes in the numerical values of Ib and IIb, they will be expressed as a decrease in the total each value of Ib and IIb. Here, this reduction rate (rate of disintegration of the peasant class) is assumed to be 20% per year at first. We also assume that the decline occurs in equal proportions between peasants' means of production and their labor-power. Accordingly, the (X+1) th year of Ib and IIb takes the following form.

Department Ib and IIb : (X+1) th year of the department of peasant production of production-goods and of consumer goods

Ib　0.8Kc+0.8Lv=S

IIb　0.8Nc+0.8Pv=T

Moreover, if the decomposition rate of peasant class is more generally expressed by p, the formulas become as follows.

Department Ib and b : (X+1) th year of the department of peasant production of production-goods and of consumer goods (part 2)

Ib $[(1-p)K]c+[(1-p)L]v=S$

IIb $[(1-p)N]c+[(1-p)P]v=T$

⑫ What has not been considered so far is IIa; the department of capitalist production of consumer goods. This department has its own difficulties, as we will see below. What I mentioned earlier in the 2nd example of Marx's schema of extended reproduction is as follows. The constant capital (c) of the 2nd department is calculated by adding the increase in variable capital (v) due to accumulation in the 1st department to the value derived from the previous year's equation for the equilibrium condition with the 1st department. The equation for equilibrium condition for the previous year was $I(v+m/2)=IIc$ in the Marx's schema of extended reproduction, but when expressed using the symbols used in this chapter, it becomes $I[B+(1-r)C]=IIE$. Also, of the capital accumulation in the 1st department, the equivalent of variable capital is $r(1-s_1)C$. In the (X+1) th year of the 2nd department, constant capital equivalent to this value is accumulated.

The issue here is peasants' production and their consumption. Since also peasants in the department of peasant production of production-goods purchase consumption means and consume them, they must purchase such consumer goods from the 2nd department. Therefore, the 2nd department must produce not only the means of consumption for the capitalists and workers of the 1st department, but also the means of consumption for the peasants of the 1st department. Consequently, an addition corresponding to consumption of this magnitude must be made to the constant capital of the 2nd department. This is calculated by subtracting the part of the disintegration of the peasant class that will proceed by the next year from the self-labor wage (L) of peasants in the department of peasant production of production-goods (Ib) in year X (i.e., $(1\text{-}p)$ L). In reality, this part may be purchased from enterprises of 2nd department or from peasants of 2nd department. However, in the reproduction schema for (X+1) th year, it is added to department IIa, not department IIb. This is because the department of IIb is a peasant class whose total product value decreases as it decomposes, so this increase cannot be added to the equivalent of constant capital of IIb. On the other hand, the 2nd department also includes not only capitalist enterprises but also peasants. The equivalent of constant capital of the latter's products is also allocated to the means of consumption of the 1st department's capitalists, workers, and peasants. Therefore, this part, that is, $(1\text{-}p)$ N, constitutes the deduction from the department IIa in the reproduction schema for year (X + 1).

To summarize the above, the constant capital (c) of department IIa in year (X+1) can be expressed as follows.

Constant capital (c) of department IIa in year (X+1) : B + $(1\text{-}rs_1)$ C + $(1\text{-}p)$ L- $(1\text{-}p)$ N

⑬ Also, the size of variable capital (v) in department IIa is $(1/s_2\text{-}1)$ times the constant capital under the condition that the capital composition, that is, c:v, is $s_2 : (1\text{-} s_2)$. Furthermore, the rate of surplus-value is 100% by prerequisite.

⑭ Based on the above conditions, the entire formula for year (X+1) of the department IIa is as follows.

Department IIa : (X+1) th year of the department of capitalist production of consumer goods

$[B+(1-rs_1)C+(1-p)L-(1-p)N]c+\{[B+(1-rs_1)C+(1-p)L-(1-p)N](1/s_2-1)\}v+\{[B+(1-rs_1)C+(1-p)L-(1-p)N](1/s_2-1)\}m=U$

⑮ Above, I have shown the general formulas for all departments in year (X+1), so let me show them all at once.

(X+1) th year

Ia $(A+rs_1C)c+[B+r(1-s_1)C]v+[B+r_1-s_1)C]m=R$

Ib $[(1-p)K]c+[(1-p)L]v=S$

IIa $[B+(1-rs_1)C+(1-p)L-(1-p)N]c+\{[B+(1-rs_1)C+(1-p)L-(1-p)N](1/s_2-1)\}v+\{[B+(1-rs_1)C+(1-p)L-(1-p)N](1/s_2-1)\}m=U$

IIb $[(1-p)N]c+[(1-p)P]v=T$

Table 7-1 was created using these general formulas. This table shows the following; the schema of extended reproduction that incorporates the "non-capitalist surroundings" is a schema that expresses the long-term historical dynamic process of a capitalist society that develops while eroding the "non-capitalist surroundings." Let me explain this below.

(1) This schema initially assumes that there will be no increase of the capital composition due to capital accumulation. Therefore, the labor supplied as a result of the disintegration of the peasant class is more rapidly absorbed into non-agricultural capital than in the case of a raised capital composition. As a result, in the 4th year, the cumulative surplus-population becomes negative and a labor shortage occurs. Until the 4th year, because the proportion of peasants in the working population is still high, the disintegration of the peasant class provides an abundant supply of low-wage labor to the capitalist sector. The contradiction (impossibility) of the commodification of labor-power does not surface because the "non-capitalist environment" is used as a source

of low-wage labor-power. This can be considered a period of primitive accumulation.

(2) This shortage of workers at the end of the period of primitive accumulation will cause wages to soar. From then on, the contradiction (impossibility) of commodifying labor-power comes to be resolved within the framework of the logic of capital through the adoption of labor-saving techniques. In my schema, this is expressed in the form of an increase of capital composition for the newly accumulated portion of capital. As a result of this increase of the capital composition, the presence of surplus-population in the labor market continues after the 4th year (part 2), but as a result of rapid capital accumulation, a turning point occurs again in the 9th year. In other words, in this year, the cumulated surplus-population will be exhausted and society will once again be faced with a labor shortage. The period from the 4th to the 9th year is when the formation of a relative surplus-population due to the increase of the capital composition is effective in alleviating the dwindling supply of labor. This is considered to be the period of formation of relative surplus-population. It should be noted that the subject here is long-term dynamic changes, and as a result, the business cycle is abstracted, and the phase of expansion of relative surplus-population due to recession is also abstracted.

(3) What is assumed from the 9th year (part 2) onwards is that, on the one hand, in order to deal with the labor shortage, the state will implement "agricultural structural policy = policy to promote the decomposition of peasants" against the background of social improvement of agricultural productivity. On the other hand, peasants are suffering from *schere* due to the huge-scaled non-agricultural capital. From the above, the rate of decomposition of the peasant class will accelerate from 20% to 50%. As a result, the labor shortage in the 9th year is avoided. However, from the 11th year onwards, it is

Table 7-1 Schema of extended reproduction incorporating "non-capitalist surroundings"

(1) Period of primitive accumulation

1st year

I a	4,000 c	+	1,000 v	+	1,000 m	=	6,000	The proportion of the peasant sector in the national economy is 27%.
I b	500 c	+	500 v			=	1,000	70% of Iam is accumulated.
II a	1,000 c	+	500 v	+	500 m	=	2,000	Iam is accumulated at a ratio of c:v=4:1.
II b	1,000 c	+	1,000 v			=	2,000	IIam is accumulated at a ratio of c:v=2:1.

The decomposition rate of peasant class in Ib and IIb is 20%.

2nd year

I a	4,560 c	+	1,140 v	+	1,140 m	=	6,840	The equation for the equilibrium condition of the 1st and 2nd departments is 1,840.
I b	400 c	+	400 v			=	800	The proportion of the peasant sector in the national economy is 21%.
II a	1,040 c	+	520 v	+	520 m	=	2,080	Labor supply from Ib, II b is 300.
II b	800 c	+	800 v			=	1,600	New labor demand for Ia and IIa is 160. New surplus-population is 140.

3rd year

I a	5,198 c	+	1,300 v	+	1,300 m	=	7,798	The equation for the equilibrium condition of the 1st and 2nd departments is 1,962.
I b	320 c	+	320 v			=	640	The proportion of the peasant sector in the national economy is 16%.
II a	1,322 c	+	661 v	+	661 m	=	2,643	Labor supply from Ib, II b is 240. New surplus-population is -60.
II b	640 c	+	640 v			=	1,280	New labor demand for Ia and IIa is 300. Cumulative surplus-population is 80.

4th year (Part 1: Capital composition remains unchanged)

I a	5,926 c	+	1,482 v	+	1,482 m	=	8,889	The equation for the equilibrium condition of the 1st and 2nd departments is 2,127.
I b	256 c	+	256 v			=	512	The proportion of the peasant sector in the national economy is 11%.
II a	1,615 c	+	808 v	+	808 m	=	3,231	Labor supply from Ib, II b is 192. New surplus-population is -137.
II b	512 c	+	512 v			=	1,024	New labor demand for Ia and IIa is 329. Cumulative surplus-population is -57.

(2) Period of relative surplus-population

4th year (Part 2: Capital composition becomes increased. For the newly accumulated part, c:v of Ia is 6:1, c:v of IIa is 3:1.)

I a	5,978 c	+	1,430 v	+	1,430 m	=	8,837	The equation for the equilibrium condition of the 1st and 2nd departments is 2,075.
I b	256 c	+	256 v			=	512	The proportion of the peasant sector in the national economy is 11%.
II a	1,563 c	+	741 v	+	741 m	=	3,046	Labor supply from Ib, II b is 192. New surplus-population is -19.
II b	512 c	+	512 v			=	1,024	New labor demand for Ia and IIa is 211. Cumulative surplus-population is 61.

5th year (Capital composition becomes more increased than the previous year. For the newly accumulated part, c:v of Ia is 8:1, c:v of IIa is 4:1.)

I a	6,868 c	+	1,541 v	+	1,541 m	=	9,949	The equation for the equilibrium condition of the 1st and 2nd departments is 2,174.
I b	205 c	+	205 v			=	410	The proportion of the peasant sector in the national economy is 8%.
II a	1,765 c	+	792 v	+	792 m	=	3,348	Labor supply from Ib, II b is 154. New surplus-population is -8.
II b	410 c	+	410 v			=	819	New labor demand for Ia and IIa is 162. Cumulative surplus-population is 53.

6th year (Capital composition becomes more increased than the previous year. For the newly accumulated part, c:v of Ia is 10:1, c:v of IIa is 5:1.)

I a	7,848 c	+	1,639 v	+	1,639 m	=	11,126	The equation for the equilibrium condition of the 1st and 2nd departments is 2,265.
I b	164 c	+	164 v			=	328	The proportion of the peasant sector in the national economy is 6%.
II a	1,937 c	+	826 v	+	826 m	=	3,590	Labor supply from Ib, II b is 123. New surplus-population is -10.
II b	328 c	+	328 v			=	655	New labor demand for Ia and IIa is 133. Cumulative surplus-population is 43.

7th year (Capital composition becomes more increased than the previous year. For the newly accumulated part, c:v of Ia is 12:1, c:v of IIa is 6:1.)

Dept.	c	v	m	=	Total	
I a	8,907 c +	1,727 v +	1,727 m	=	12,361	The equation for the equilibrium condition of the 1st and 2nd departments is 2,350.
I b	131 c +	131 v		=	262	The proportion of the peasant sector in the national economy is 5%.
II a	2,088 c +	851 v +	851 m	=	3,790	Labor supply from Ib, II b is 98. New surplus-population is -15.
II b	262 c +	262 v		=	524	New labor demand for Ia and IIa is 113. Cumulative surplus-population is 28.

8th year (Capital composition becomes more increased than the previous year. For the newly accumulated part, c:v of Ia is 14:1, c:v of IIa is 7:1.)

Dept.	c	v	m	=	Total	
I a	10,035 c +	1,808 v +	1,808 m	=	13,651	The equation for the equilibrium condition of the 1st and 2nd departments is 2,431.
I b	105 c +	105 v		=	210	The proportion of the peasant sector in the national economy is 3%.
II a	2,221 c +	870 v +	870 m	=	3,962	Labor supply from Ib, II b is 79. New surplus-population is -21.
II b	210 c +	210 v		=	419	New labor demand for Ia and IIa is 100. Cumulative surplus-population is 7.

9th year (Part 1: Capital composition becomes more increased than the previous year. For the newly accumulated part, c:v of Ia is 16:1, c:v of IIa is 8:1.)

Dept.	c	v	m	=	Total	
I a	11,226 c +	1,882 v +	1,882 m	=	14,990	The equation for the equilibrium condition of the 1st and 2nd departments is 2,508.
I b	84 c +	84 v		=	168	The proportion of the peasant sector in the national economy is 3%.
II a	2,340 c +	885 v +	885 m	=	4,111	Labor supply from Ib, II b is 63. New surplus-population is -26.
II b	168 c +	168 v		=	336	New labor demand for Ia and IIa is 89. Cumulative surplus-population is -19.

(3) Period of modern capitalism

9th year (Part 2: Rate of decomposition of the peasant class accelerates to 50%.)

Dept.	c	v	m	=	Total	
I a	11,226 c +	1,882 v +	1,882 m	=	14,990	The equation for the equilibrium condition of the 1st and 2nd departments is 2,477.
I b	52 c +	52 v		=	105	The proportion of the peasant sector in the national economy is 2%.
II a	2,313 c +	882 v +	882 m	=	4,077	Labor supply from Ib, II b is 275. New surplus-population is -24.
II b	164 c +	164 v		=	328	New labor demand for Ia and IIa is 299. Cumulative surplus-population is 20.

10th year (50% of the rate of peasant disintegration continues. The capital composition is even more advanced than the previous year.
For the new accumulation part, c:v of Ia is 18:1, c:v of IIa is 9:1.)

Dept.	c	v	m	=	Total	
I a	12,474 c +	1,951 v +	1,951 m	=	16,377	The equation for the equilibrium condition of the 1st and 2nd departments is 2,542.
I b	26 c +	26 v		=	52	The proportion of the peasant sector in the national economy is 1%.
II a	2,460 c +	898 v +	898 m	=	4,257	Labor supply from Ib, II b is 108. New surplus-population is 22.
II b	82 c +	82 v		=	164	New labor demand for Ia and IIa is 86. Cumulative surplus-population is 42.

11th year (50% of the rate of disintegration of the peasant class continues. The capital composition is stagnant.
The composition of new accumulation part remains the same as the previous year at c:v = 18:1 for Ia and c:v = 9:1 for II a.)

Dept.	c	v	m	=	Total	
I a	13,768 c +	2,023 v +	2,023 m	=	17,815	The equation for the equilibrium condition of the 1st and 2nd departments is 2,622.
I b	13 c +	13 v		=	26	The proportion of the peasant sector in the national economy is 0%.
II a	2,581 c +	912 v +	912 m	=	4,404	Labor supply from Ib, II b is 54. New surplus-population is -31.
II b	41 c +	41 v		=	82	New labor demand for Ia and IIa is 85. Cumulative surplus-population is 11.

12th year (50% of the rate of disintegration of the peasant class continues. The capital composition is stagnant.
The composition of new accumulation part remains the same as the previous year at c:v = 18:1 for Ia and c:v = 9:1 for II a.)

Dept.	c	v	m	=	Total	
I a	15,110 c +	2,098 v +	2,098 m	=	19,306	The equation for the equilibrium condition of the 1st and 2nd departments is 2,711.
I b	7 c +	7 v		=	13	The proportion of the peasant sector in the national economy is 0%.
II a	2,691 c +	924 v +	924 m	=	4,539	Labor supply from Ib, II b is 27. New surplus-population is -60.
II b	20 c +	20 v		=	41	New labor demand for Ia and IIa is 87. Cumulative surplus-population is -49.
	→Dependence on external labor force (introduction of immigrants, foreign direct investment)					

Note 1) Equation for the equilibrium condition of the 1st department = (Current year Ia) v + (Current year Ib) v + (Previous year I) m × 0.3
Equation for the equilibrium condition of the 2nd department = (Current year IIa) c + (Current year IIb) c

2) Proportion of the peasant sector in the national economy = Agricultural production value / Total production value × 100 = (Ib + II b) / (Ia + Ib + II a + II b) × 100

assumed that constant capital has already grown so large that capitalist companies will become reluctant to adopt new technologies, and as a result, the capital composition will become stagnant.[163] Although the acceleration of the decomposition of the peasantry contributes to the supply of new labor to the capitalist sector, the stagnation of the capital composition creates a situation in which labor is rapidly absorbed by the capitalist sector. The period in which these various factors intertwined in a complex manner will be referred to here as the period of modern capitalist society.

(4) After going through these processes, in the 12th year, the cumulative surplus-population becomes negative for the third time, and the society is once again in an era of worker shortage. At this point, there is nothing left within this society to procure workers. Enlarged capital is reluctant to adopt labor-saving technologies.[164] The peasant class has already been completely disintegrated and can no longer serve as an abundant source of labor for the capitalist sector. Therefore, if the capitalist sector of society attempts to further accumulate, it will be forced to seek workers from outside society. On the one hand, this means promoting the immigration of foreign workers into the country, and on the other hand, companies making foreign direct investments

---

163 "In the early 1970s, developed countries experienced an over-accumulation of capital that exceeded the supply of labor. Labor shortages became more serious, and real wages rose significantly. The resulting crisis in capital accumulation accompanied by a decline in profits was reversed through IT rationalization, and a process of large-scale reshaping of the relative surplus-population favorable to capital developed over a long period of time." (Ito 2018) Even if this were true, wouldn't the half-century-long process of introducing labor-saving technologies to alleviate labor shortages still be described as "stagnant?"

164 It is necessary to pay close attention to the impact of technological innovations brought about by ICT and AI. There are also the following prospects associated with these technologies. "The development of ICT and AI technologies is providing the technological foundation for eliminating the instability of the market economy and realizing a planned economy that balances supply and demand." (Fujita 2021)

and hiring new workers in the investment destinations.[165]

6. Conclusion—Perspective from the "Shema Incorporating Non-Capitalist Surroundings"

In this chapter, I examined the Rosa-Luxemburg's schema of extended reproduction, and then proposed a schema of extended reproduction that incorporates the "non-capitalist surroundings." Through this work, the following can be said.

① The necessity of "non-capitalist surroundings" for capitalist production is difficult to derive from a sales perspective, as Luxemburg tried to do by making full use of the schema of extended reproduction. However, as this chapter has done, it can be derived smoothly from a labor market perspective.

② This perspective stems from the fact that the basic contradiction of capitalist society is seen as the contradiction (impossibility) of the commodification of labor-power. In the end, capitalist society cannot resolve this contradiction (impossibility) within the capital-labor relationship, so it attempts to resolve the contradiction relying on the "non-capitalist surroundings." Luxemburg saw the basic contradiction of capitalist society in the imbalance between departments, and therefore tried to explain the necessity of the "non-capitalist surroundings" from the perspective of sales channels.

③ The schema of extended reproduction that incorporates the "non-capitalist surroundings" expresses the long-term historical dynamic process of capitalist production, which accumulates while eroding the "non-capitalist

---

165 In addition to the disintegration of the peasantry in developing countries, it is also necessary to pay close attention to the effects of the population explosion there.

surroundings." Therefore, this schema can represent the following series of life events in capitalist society; primitive accumulation, raise of capital composition, cumulative increase in relative surplus-population and its depletion, improvement of agricultural productivity and agricultural structural policy, technological perishability under the enormous growth of constant capital, external dependence of capitalist production (or the downfall of this mode of production). This schema makes it possible to express that this mode of production is a body in motion that simultaneously undermines its own foundations of existence as it grows and develops, eventually heading toward collapse.

Final Chapter: Historical View from Primitive Accumulation

1. Theme

In the final chapter, I propose a historical perspective from primitive accumulation for understanding the historical development process of modern capitalist society. Primitive accumulation is a process in which direct producers, mainly peasants, who were previously tied to the means of production (mainly land), are historically separated from the means of production. The workers created in this way are absorbed into the rising industrial capital in search of a chance to survive. Therefore, primitive accumulation is at the same time a process of historical creation of capital-labor relations (capital relations). [166]

166 Regarding the theory of primitive accumulation, Baba (1986) states; "I don't actually know the extent to which the theory of primitive accumulation has been developed since 'Das Kapital', or perhaps it has been neglected and stagnated." Looking at the debates in Japan that I know of, first of all there is the debate between the Otsuka school and the Uno school. This debate is over whether the main aspect of primitive accumulation is a "peaceful" process based on competition between producers, or whether it is a process in which violence is the main element. According to the former, the "bipolar decomposition" of "middle producers," that is, independent farmers (yeomanry) and small masters of rural industries expanding outside privileged cities, through competition in their productivity, gave rise to industrial capitalists and workers. (Otsuka 1956a) The Uno School, on the other hand, regards the early accumulation of merchant capital and usurious capital as the primary element in capital formation, and when it comes to creating a proletariat, it emphasizes the significance of violence when peasants are deprived of their land. (Uno 1971) Next, Mochizuki (1977, 2000) argues that Marx's "Forms" (1858) should be read as an introduction to the theory of primitive accumulation, through an examination of the 1857-1858 draft "*Grundrisse.*" He raises questions about the relationship between these two classics. Furthermore, Mochizuki (1981b, 1982) organizes the concept of primitive accumulation after reading "*Das Kapital*" Volume 1, Chapter 24, "the so-called primitive accumulation." Regarding the differences between his understanding of the concept of primitive accumulation

The historical view from primitive accumulation proposed in this chapter understands the history of capitalist society as a process of primitive accumulation, shifting the focus from the capital side, which has been emphasized in many conventional theories, to the labor supply side. Primitive accumulation is the above-mentioned "relationship" between the capital sector and the non-capital sector. Therefore, the historical view from primitive accumulation is a way of understanding the history of capitalist society that focuses on the articulation between the two.

Furthermore, the concept of primitive accumulation has traditionally been associated with the image of "the starting point of capitalist society." In contrast, what I propose here is to understand the entire history of capitalist society as one primitive accumulation process. This is a process in which the area of surrounding non-capitalism, which is subject to digestion and metabolism by capital, gradually expands, and eventually the non-capitalist area may be completely consumed. This digestive process first began in rural England, and over time it is now being developed in rural areas in developing countries. There appears to be a distance between my vision of primitive accumulation and the general vision of primitive accumulation.[167] Therefore, the

---

and mine, see notes 168 and 170 in this chapter. Third, Fujise (1985) focuses on Marx's (1881) following recognition: Depending on whether land ownership in a pre-capitalist society was private or communal, the types of capitalist societies that were formed later differed. He also explains the necessity of understanding the various types of primitive accumulation. I organize the concept of primitive accumulation in relation to capital accumulation theory and community theory (Chapters 1 to 4 in this book), and discuss various types of primitive accumulation (Chapter 6, Final chapter). In addition, in Europe and the United States, there is a tendency, led by world system theorists, to apply the term primitive accumulation to the transfer of value from the non-capitalist sector and housewives to the capitalist sector. (Amin 1970, Meillassoux 1975, Mies et al. 1995)

167 This chapter also attempts to provide an answer to the following question posed by Mochizuki (1981b), while independently rearranging the concept of primitive

proposal for the historical view from primitive accumulation must be to bridge this distance.

In any case, the first thing that needs to be done is to present the concept of primitive accumulation that I believe. However, this has already been discussed in the previous chapters of this book (Chapters 1-3). Therefore, in the next section, I will summarize the contents of these chapters, although I will add some points. In Section 3, I propose the historical view itself from primitive accumulation. Section 4 is a typology of each country's capitalist societies from the historical view from primitive accumulation, and Section 5 is a proposal to classify the stages of development of world capitalism. This chapter is the result of research conducted after the publication of Yamazaki (2014b). Although this chapter incorporates some of essence of Yamazaki (2014b), it attempts to launch a completely new concept of the historical view from primitive accumulation.

2. What Is Primitive Accumulation?

Primitive accumulation is a process in which direct producers, mainly peasants, who were previously connected to the means of production, are historically separated from the means of production (land). [168] However, if we

---

accumulation. "Did primitive accumulation occur only once in world history? Is it once for each nation state? Is it something that is being carried out on a daily basis, parallel to the accumulation proper of capital? Marxism now stands before these three problems, without questioning the content of the term 'primitive accumulation.'"

168 "The so-called primitive accumulation, therefore, is nothing else than the historical process of divorcing the producer from the means of production." (Marx 1867) "Nothing else", or in the German version "*nichts als*", is a strong expression. This sentence can be read as "primitive accumulation = the historical separation process between producers and the means of production," and "primitive accumulation ≠ other social phenomena." I believe that the concept of primitive accumulation should be centered around this proposition in Chapter 24, Section 1 of Capital, Volume 1. (Yamazaki 2019, Chapter 3 of this book) Furthermore, Mochizuki

look at the process in more detail, it consists of the following two components. The first is the component by which the communities that traditionally protected the direct producers are broken up through legal and illegal means, leading to the establishment of modern land ownership. The second is the component by which direct producers break down and many of them become workers.

Furthermore, if we look at the process consisting of these two components as a whole, we can see it as a process in which the "principle of survival" that characterized pre-capitalist communal society is denied, and furthermore, it is ultimately destroyed. The "principle of survival" here is defined as the purpose of economic activity in a pre-capitalist communal society. The purpose of economic activity there is to produce and secure products such as agricultural products in order to sufficiently guarantee the survival of the individuals who make up society and their reproduction beyond generations.[169] This is a very different form of society from a capitalist society where the purpose of economic activity is to maximize corporate profits. Therefore, in the latter case,

---

(1982) defines this as "a standard proposition that everyone remembers," and then independently presents an understanding of primitive accumulation centered on the description in Chapter 24, Section 6. According to him, primitive accumulation consists of "the primitive accumulation of funds" and "the primitive accumulation of labor-power." In my understanding, the so-called "primitive accumulation of funds" is one of the "elements of primitive accumulation" that form the background of primitive accumulation during the "period of primitive accumulation." It is distinct from the primitive accumulation itself. I sympathize with Professor Mochizuki's problem that "a rigorous understanding of Marx's theory of primitive accumulation is essential." However, I think it is important to consider the consistency between the "standard proposition" and the other descriptions in Chapter 24.

169 "in all these forms (Asian, Roman, and Germanic forms of pre-capitalist ownership; citer), where landed property and agriculture form the basis of the economic order, and consequently the economic object is the production of use-values – i.e., the *reproduction of the individual* in certain definite relationships to his community, of which it forms the basis" (Marx 1858)

the goal is not necessarily the survival of the people who make up society.

By the way, we need to reconsider what kind of situation it is when direct producers, mainly farmers, are "historically" separated from the means of production (land). This is because it goes without saying that primitive accumulation cannot be said to have been accomplished just because a very small number of direct producers have lost their means of production (land). Conversely, it would also be an extreme assumption to say that primitive accumulation cannot be considered accomplished unless all direct producers in the country lose their means of production (land). [170] In fact, there is a threshold between these two extremes, and when the state of society crosses that threshold, we can mark it as the accomplishment of primitive accumulation. Where is this threshold? My answer to this question is to understand this concept (primitive accumulation) as the process of creating the preconditions for the law of relative surplus-population to function, that is, the process of preparing the start of this law.[171] In other words, when the capitalist law of population begins to operate, we should consider that primitive accumulation has already been accomplished and is a thing of the past.

This is because I believe that the following is necessary condition for the

170 The idea of the "accomplishment" of primitive accumulation also exists in *Das Kapital*, as seen in the following quotation. "In Western Europe, the home of Political Economy, the process of primitive accumulation is more or less accomplished." (Marx 1867) Furthermore, even after the "accomplishment" of primitive accumulation, the separation of direct producers and means of production continues. Mochizuki (1982) calls this situation "additional primitive accumulation," and rather considers it as a continuation of the primitive accumulation.

171 Does Marx's law of relative surplus-population address the question of the mechanism that reduces the wage rate within the limits compatible with the growth of capital's value during the business cycle? Or does it discuss the theory of trends beyond cycles, the law of trends, the inevitability of progressive increase of industrial reserve army? There are these discussions. Regarding this controversy, see Koga (1977) and Arai (1985). I believe in the former of these, based on the theory of business cycle.

mechanism of labor force creation through the increase of the organic composition of capital to be put into motion. That is, as a result of the decomposition of the non-capitalist sector centered on farmers, the state in which low-wage labor is abundantly supplied to the capitalist sector will no longer exist. In other words, because sources of low-wage labor in the non-capitalist sector have already been exhausted, capital is forced into a corner where it becomes necessary to create its own labor procurement mechanism through the increase of its organic composition. I believe that this is a precondition for the capitalist law of population to begin operating. From then on, the capitalist law of population came to determine the dynamics of the business cycle. On the other hand, in societies where the non-capitalist sector plays a major role in supplying labor to the capitalist sector, the wages paid to workers by the capitalist sector are kept low. In such a society, the capitalist sector is able to impose a significant portion of the labor reproduction costs on the non-capitalist sector, which it would normally have to bear. (Meillassoux 1975). This allows the capitalist sector to escape its share of social responsibility. Moreover, the surplus-population pressure created by the labor supplied from non-capitalist sector makes such low wages a reality in the labor market. Furthermore, during the period in which the capitalist sector can employ the labor force abundantly supplied by the non-capitalist sector at low wages, there is a limit to the introduction of labor-saving technologies from a cost-effectiveness perspective.[172] As a result, the operation of the capitalist system of labor procurement, which is based on the creation of a relative surplus-population, will be suppressed. In this way, the theory of primitive accumulation is theoretically understood as a theory of the process of creating the precondition for the law of relative surplus-population to begin to operate. This understanding differs from the conventional understanding, which is valid

172 Yamazaki's Collected Works Vol. 2, Monthly Report, pp.5-8.

in itself, which views the historical separation process between direct producers and the means of production such as land as an essential element of the primitive accumulation process. My understanding asserts its uniqueness in understanding primitive accumulation in that it emphasizes the relationship between primitive accumulation and the law of relative surplus-population.[173] However, things do not proceed in this pure form, as latecomers to capitalist production are able to introduce advanced technology from advanced countries. In this case, the organic composition of capital increases gradually without being urged by the inherent factor of sufficient disintegration of the non-capitalist sector.

Furthermore, a society in the period of primitive accumulation becomes unstable. That's natural. Direct producers such as peasants constituted the majority in pre-capitalist society, and at the same time had a stable status as they were firmly tied to the means of production (land) through communal customs. They will lose their means of production and will be forced to become unstable proletarians. Moreover, since primitive accumulation destroys the community and the social order of neighborliness and mutual aid that governs the community, it provokes legitimate criticism among the people. Therefore,

173 Regarding the relationship between relative surplus-population and primitive accumulation, the following is written in Volume 1 of "Capital." "The organization of the capitalist process of production, once fully developed, breaks down all resistance. The constant generation of a relative surplus-population keeps the law of supply and demand of labor, and therefore keeps wages, in a rut that corresponds with the wants of capital. The dull compulsion of economic relations completes the subjection of the laborer to the capital. (Omission) It is otherwise during the historic genesis of capitalist production. The bourgeoisie, at its rise, wants and uses the power of the state to 'regulate' wages, i.e., to force them within the limits suitable for surplus-value making, to lengthen the working-day and to keep the laborer himself in the normal degree of dependence. This is an essential element of the so-called primitive accumulation." Here, we can discern the flow of the logic of "primitive accumulation (historical nascent period of capitalist production) → law of relative surplus-population."

societies in the period of primitive accumulation, especially in the early stages of capitalist development without social security, are subject to revolutionary social upheavals. There are many examples of this in past history, such as the 17th century English Revolution (Puritan Revolution and Glorious Revolution), the French Revolution (1789-1799), the Russian Revolution (1917), and the Iranian Revolution (1979). [174]

3. Historical View from Primitive Accumulation

Here, I summarize the historical view from primitive accumulation in the following propositions. One is to divide the history of a country's capitalist society into stages diachronically into the period of primitive accumulation and the period of post-primitive accumulation (**Table F-1**). The period of primitive accumulation is the period from the beginning of primitive accumulation (the beginning of the destruction of the community) to its end (the beginning of the capitalist law of population) in the sense described in the previous section. During this period, the outflow of labor from the non-capitalist sector through its disintegration became the main source of low-wage labor for the capitalist

174 See Note 194 for the French Revolution and Note 202 for the Iranian Revolution. The social background of the British Revolution was the disintegration of the peasantry and the "enclosure" movement (a revolt to break down the "enclosure" as a reaction). The Levellers, who were in conflict with the landowning parliamentarians within the revolutionaries who opposed the royalists, dreamed of a society of simple commodity production while attacking a certain degree of bourgeois development. For more information on the British Revolution, see Horie (1962). Furthermore, Yasuda (1971: Chapter 3) viewed the Russian Revolution from the aspect of the revival movement of the Mir community in rural areas, which corresponded to the movement of urban workers. According to Ominami (1986), the "vulnerability of *régulation*" in developing countries results from their "chaotic social relationships." "It has become extremely difficult for various social sectors to overcome the narrow framework of corporatism and express universal interests. Society and economy are thus forced to function less by consensus than by coercion through open violence."

industrial sector. On the other hand, the post-primitive accumulation period is the period after primitive accumulation has been accomplished and the capitalist law of population begins to operate. With this law in place, the capitalist sector acquires its own labor procurement mechanism inherent in accumulation. Here, primitive accumulation is viewed as a national process, that is, as one stage in the history of the development of a country's capitalist society.

On the other hand, however, it is also possible to relate primitive accumulation to the world-historical development process of capitalist society, as we will see next.

Primitive accumulation is a social phenomenon that occurs only once in history in a certain region, regardless of the length of the period. On the other hand, primitive accumulation does not necessarily proceed simultaneously in each region of the world, that is, across regions across the globe. Historically, primitive accumulation first began in Western Europe, with Britain as the central region, and then spread successively to other parts of the world. In this way, the primitive accumulation that constantly spreads from the central region to the surrounding regions and the capital relations that are generated as a result of the primitive accumulation, have been summarized by country in the process of their development. These have emerged, for example, in the form of "British primitive accumulation" , "British capitalism" , "Japanese primitive accumulation" , and "Japanese capitalism" , using the name of each country as an adjective. (Otsuka 1960) If we emphasize this aspect of "summarizing on a country-by-country" basis, we can view primitive accumulation as a national process as described above. However, if we focus on the aspect that the primitive accumulation started in England and gradually spread to surrounding areas, then the primitive accumulation is a one-time historical event. Moreover, as I have shown on another occasion, even in the

Table F-1 Periodization of primitive accumulation

Pre-primitive accumulation period	Primitive accumulation period	
	Fracture of community = beginning stage of primitive accumulation	
	↑	↑
Commencement of usurpation of common land		Transition of society from feudalism to capitalism =Establishing modern land ownership = Fulfillment of the 1st component of primitive accumulation

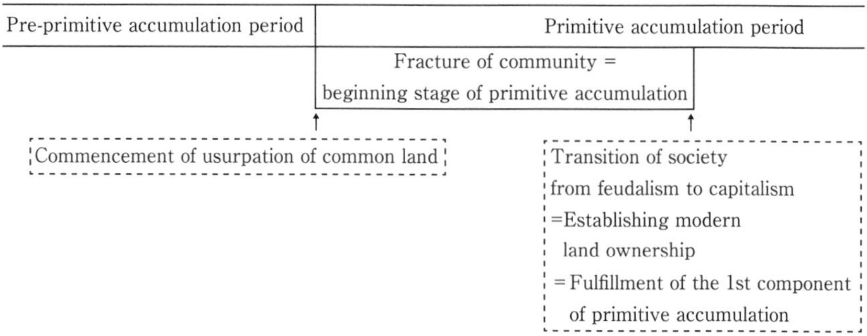

21st century, the primitive accumulation is still ongoing in developing countries as a primitive accumulation inherent in the "periphery," (Fujise 1985, Chapter 6 of this book) so it has not yet been completed. A capitalist society that appeared in history as the "European world economy" during the "long 16th century" from the late 15th century to the first half of the 17th century, (Wallerstein 1974) and today has grown into a unified entity that operates on a global scale is expressed as the "world system" (ibid.) or "world capitalism." (Iwata 1964) From the above perspective, it is not far-fetched to say that the development of the "world system" or "world capitalism" has always coexisted with primitive accumulation. In other words, the "world system" or "world capitalism" has had primitive accumulation as one of its components. (Luxemburg 1913)

Should we be satisfied with such a combination of both theories? Or should we delve deeper into the question of whether primitive accumulation is a national process, or whether it is something that pervades the history of the global expansion of capitalist society? This question is connected to another question: Should capitalist society be seen as an entity summed up in a single country, or should it be seen as an organic whole within a global context?

Here, I would like to think about this question as follows. First, primitive

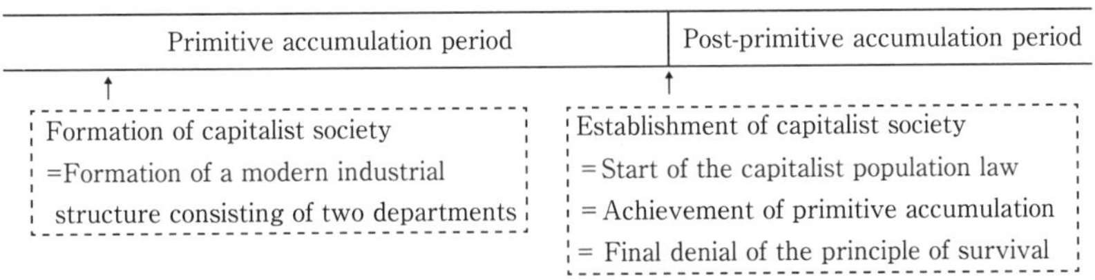

Primitive accumulation period	Post-primitive accumulation period
↑	↑
Formation of capitalist society =Formation of a modern industrial structure consisting of two departments	Establishment of capitalist society = Start of the capitalist population law = Achievement of primitive accumulation = Final denial of the principle of survival

accumulation requires the existence of modern land ownership in society in order for its results to be permanent. Suppose that in a region where modern land ownership has not been established, direct producers, mainly peasants, were separated from the means of production (land) for some reason. In that case, they cannot bear the economic pain of not having the means of production, so they move into unoccupied land without an owner, occupy it, and resume food production there. Even if that were the case, the law would not be able to prohibit them from doing so.[175] Therefore, in order to make the results of primitive accumulation permanent, it is a prerequisite to abolish customary communal land ownership within a certain extent of land, that is, within a spatial area. In addition, there must be a national law that officially recognizes modern private land ownership. Therefore, primitive accumulation presupposes the existence of a modern state that enacts such a law. Furthermore, in reality, primitive accumulation has not progressed simultaneously throughout the world. It has progressed with a before-and-after relationship between advanced and

175 I have discussed the following point on another occasion based on a field survey. (Yamazaki 2007, Collected Works Vol. 4) The "occupation of unoccupied land" was actually carried out until very recently in the Niger River inland delta region, where communal land ownership is practiced. Yoshida (1975) provides a systematic description of Africa's pre-modern land system.

latecomer regions. Therefore, primitive accumulation has proceeded as a national process within each region, with time lags between regions. Capitalist society, which is born through primitive accumulation, has been generally summarized as "capitalism in one country." [176] On the other hand, as mentioned earlier, the commodity economy constantly infiltrated from the core region of capital relations to the peripheral regions where non-capitalist relations were dominant, and this caused a chain reaction of primitive accumulation in the latter. Therefore, while primitive accumulation is, on the one hand, a process that can be summarized in a single country, at the same time it has spread and chained like a domino to surrounding areas, and has ultimately developed into a global process. When focusing on the latter aspect, the "world system" or "world capitalism" is confirmed as an independent entity with actual conditions that is distinct from "capitalism in one country."

It should be noted that my argument, which emphasizes the continuous propagation of capital relations from the core region to the peripheral regions, is expected to be met with objections that emphasize the "discontinuity" of this propagation process. For example, Obata (2011) states: "When the core develops, capitalistic development in the periphery is suppressed, and as the core matures, the pressure for capitalism increases in the periphery, and intensive capitalistic development progresses." According to this discontinuity theory, the responsibility for the capitalistic development of the peripheral region, or conversely, the responsibility for its stagnation, will be primarily placed on the development trends of the core region. In contrast, the historical view from primitive accumulation focuses not only on the core region but also on the peripheral regions, and sees inside the latter an important element for the development of capital relations. The commodity economy constantly

176 I learned this way of understanding primitive accumulation from Muraoka's (1988) definition of the national economy.

permeates from the core region to the peripheral regions. When the degree of penetration reaches a certain level, it will lead to the destruction of the community. After this, the disintegration of direct producers, the transformation of many of them into workers, and the formation of capital relations will proceed all at once. By focusing on internal trends in peripheral regions in this way, it becomes possible to grasp the following aspects in a unified manner. That is, there is the aspect of "continuous change" in society as the commodity economy permeates, and the aspect of "intermittent change" in society that is triggered by the crushing of communities and leads to the development of capital relations.

Based on the above considerations, I would like to define the concept of world capitalism as follows, using the concept of primitive accumulation. Today's world capitalism is a complex polymer made up of all the societies on earth.[177] Each country and society that makes up world capitalism has its own level of development. The stages of development include pre-capitalist and pre-primitive accumulation communal societies before the onset of primitive accumulation (Sub-Saharan Africa), in addition to societies in the primitive accumulation period (developing countries excluding Sub-Saharan Africa) and post-primitive accumulation period (developed countries). [178] Considering this point, world capitalism is a complex polymer woven by national societies, each of which is in one of the stages of development; pre-primitive accumulation, primitive accumulation, or post-primitive accumulation.

At the same time, it may also be possible to consider the primitive

177 Nawa (1949) views the world economy as something in which "individual citizens form an integral part."

178 As mentioned earlier, the modern state has the function of officially recognizing modern land ownership. If this is the case, the countries of sub-Saharan Africa, which still live under customary and communal land ownership, will lack important requirements for their survival as modern states.

accumulation of world capitalism from a different perspective. As mentioned above, it first began in Western Europe, with Britain as its core region, and then gradually spread to other parts of the world. Moreover, it is not over even today. This is because in today's developing countries, on the one hand, there is active differentiation and disintegration of the peasantry class, and on the other hand, the rapid acquisition of farmland by foreign capital and the corresponding loss of land by peasants is progressing. (Takeuchi 2017) In this sense, capitalist society has historically always coexisted with primitive accumulation up to the present day.

As stated at the beginning of this chapter, the historical view from primitive accumulation attempts to understand the history of capitalist society as a single primitive accumulation process. What this means is that world capitalism has always coexisted with primitive accumulation. Therefore, when viewed across a cross-section at a particular point in time, world capitalism can be seen as a collection of national societies, each at the following stages, that are synchronized and superimposed; pre-primitive accumulation period, primitive accumulation period, and post-primitive accumulation period.

4. Types of Primitive Accumulation as a National Process

The primitive accumulation as a national process is diverse and varies from country to country,[179] but is it not possible to find typological differences within this diversity?

I will explain this typology here with reference to Wallerstein (1974). As mentioned earlier, he argues that modern capitalist society was born in an environment in which the "European world economy," which included the New World, was established during the "long 16th century." At the same time, it

179 In the French version of *Das Kapital* (1872-1875), Marx limited the validity of the book's description of primitive accumulation to Western European countries.

was also the establishment of a three-layer structure between regions; "core," "semi-periphery," and "periphery." Needless to say, since then the world economy has become much broader than the "European world economy" of the past, and today it is literally global. However, extending Wallerstein's argument recognizes the fact that there is a "complementary divergence" within the globalized world economy. Furthermore, the primitive accumulation process within a country can be largely viewed as consisting of two types of processes that are complementary to each other; a "core" process and a "peripheral" process. (Fujise 1985)

By the way, Amin (1970) develops the following argument. In the "periphery", the "pre-capitalist structure" that precedes the capitalist society there is transformed into the "peripheral capitalist structure" through the infiltration of capital relations from outside. However, even if the latter state is reached, capital relations will not completely dissolve the non-capitalist sector. Therefore, even if capital relations become dominant in society, they do not tend to become exclusive. Therefore, the heterogeneous nature of the social structure, including non-capitalist sector, stubbornly persists there.[180] In contrast, in the "core," the social structure is "similar to the pure model of *Das Kapital*, characterized by the polarization of social classes into two basic classes." While dilating on Amin's points, I would like to define the "core" process and the "peripheral" process of primitive accumulation as follows.

a) The "core" process of primitive accumulation progressed with the prospect of its completion due to the "polarization of society," and in fact, it has already been accomplished in the developed countries and has become history. On the other hand, (b) in the "peripheral" part of the primitive accumulation process of the developing countries, its future accomplishment cannot be easily

180 "This would correspond to the 'reversal phenomenon of the purifying tendency' in Uno's theory." (Motoyama 1976)

predicted because "the heterogeneous nature of the social structure stubbornly continues." Therefore, the "core" process of primitive accumulation is exclusively the subject of historical analysis in developed countries. On the other hand, the ongoing "peripheral" process is the subject of current situation analysis and historical analysis in developing countries. However, I cannot help but feel that Amin's argument and my elaboration of it are influenced by the dependency theory watching fatal stagnation in the economy of "peripheral" developing countries. In other words, the prospect that the bipolarization and decomposition of social classes into capitalists and laborers will not proceed completely in "peripheral" developing countries is together with the pessimistic view of economic development there. However, today, East Asian countries are experiencing rapid economic development since the fourth quarter of the 20th century. Therefore, it is no longer possible to exclude the possibility that there are regions within the "peripheral" developing countries where differentiation and disintegration into two basic classes is progressing thoroughly. Because of this reservation, what has been stated above is still a provisional definition of the "core" process and the "peripheral" process of primitive accumulation.

By the way, with regard to the "core" process of primitive accumulation, it seems possible to grasp both ends of the spectrum; the first-mover country model, typified by Britain, and the latecomer model, of which Japan is an example. The distinguishing feature is that in the former, the following three elements related to the born of capitalist society developed simultaneously and concurrently within a relatively short period of time, whereas in the latter, these three elements appeared dispersed over a relatively long period of time. The three elements are: ① The modernization of land ownership, that is, the "transition of society from feudalism to capitalism," which involves the establishment of private land ownership; ② "formation of a capitalist society" from the perspective of the formation of a modern industrial structure consisting of two departments due to industrialization; ③ "establishment of a capitalist society" from the perspective of invoking the law of population in a capitalist society, in which a capitalist society establishes its own labor procurement mechanism.[181] As seen in Section 2 of this chapter and as discussed in Chapter 1, I consider ① of these to be the fulfillment of the first component of primitive accumulation, which have already begun. Moreover, ③ is regarded as the completion of the primitive accumulation itself. Furthermore, a characteristic of the first-mover country type was the fact to compress the period from ① to ③ in a short time, whereas in Japan, a latecomer country, this period lasted for a long time.

In England, the usurpation of common land that began at the end of the 15th century was fulfilled with the enactment of the General Enclosure Act of 1801, which corresponds to ①.[182] The period from then until the first cyclical economic crisis in 1825, which corresponds to ③, lasted only about 25 years.[183]

---

181 The "establishment of a capitalist society" is also the completion of "the tyrannical control of capital over labor in the production process, which is the real subsumption of labor under capital." First, the trigger for capital's tyrannical control over labor in the production process is the use of machinery itself. Because machines dismantle manual skill and simplify labor, the mental element of the labor process is alienated from labor and transformed into the power of capital. At this point, a modern industrial structure has already been formed, and a capitalist society has therefore been formed. On the other hand, in order for capital's tyrannical control over continuously to be completed, it is necessary for capital's organic composition to become continuously increased, leading to a relative surplus in the worker population and for competition among workers to become the norm. The last point leads to a deepening of the subordination of labor to capital.

182 The passage of the General Enclosure Act marked the final stage in the history of enclosure. The background was the situation at the time, in which agricultural development was no longer possible without enclosure. The General Enclosure Act ultimately ended British open farmland. (Shiina 1962)

183 In the process of emerging from the depression following the panic of 1825, the organic composition of capital increased through the introduction of new production methods (the self-actor invented in 1825 as an improved mule spinning), unemployment and poverty became more serious, and the labor

In Japan, on the other hand, there was a period of about 100 years between the land tax reform of 1873, which corresponds to ①, and the 1980s, which I believe corresponds to ③.[184]

Even so, in the case of latecomer countries in the "core," the start of primitive accumulation occurred in time to achieve the subsequent bipolarization and decomposition of the social structure. However, in the "peripheral" process of primitive accumulation, enormous external pressure from "core" countries initially fixed the vertical international division of labor between primary and secondary products (*disarticulation*; Amin 1970). Furthermore, in the latter half of the 20th century, when some developing

---

movement became more radical. (Hudson 1992) In other words, the formation of the Grand National Consolidated Trades Union led by Owen (1834) and the development of the Ten-Hour Movement instigated by Oastler. (Yoshioka 1981) This was also the time when Wakefield developed his argument for "systematic colonization" in "England and America" (1833), and the colony came into the limelight as a means of alleviating surplus-population and poverty. (Yoshioka 1981)

184 In a separate paper, I argue that it was after the 1980s, a turning period, that capital accumulation in the Japanese economy began to develop solely on the basis of capitalist population law. (Yamazaki 2014a, Collected Works Vol. 2) As a latecomer, the Japanese economy pursued catch-up until the 1970s. At that time, capital accumulation relied heavily on cheap labor supplied by domestic peasants. Therefore, there was little need to internalize the mechanism of labor force creation within the movement of capital accumulation itself. However, with the 1980s as a turning period, peasants who exhausted their labor were no longer able to play the role of a source of labor outside of agriculture. As a result, the Japanese economy began to use neo-conservative policies as a lever to increase unemployment. Under this situation, Japan's economy shifted to one dependent on the unemployed. Under the unemployed-dependent model, unemployed people created during recessions are absorbed by capital during booms. In other words, the mechanism of labor force creation is internalized within the process of capital accumulation. Since the 1980s, there was a rapid increase in non-regular employment, symbolized by temporary workers, as a form of employment suited to this situation. The technological foundation that made this rapid increase possible was the simplification of labor through OA and FA. (Ito 1990)

countries achieved industrialization through foreign direct investment from the "core," foreign capital came towards bonded processing zones in order to strengthen their competitiveness in external markets. Such foreign capital generally involves an increased capital composition and does not create supporting industries within developing countries. For these reasons, the non-capitalist sector in "periphery" was not sufficiently dismantled (*extraversion*).

What are the differences between countries in the "peripheral" process of primitive accumulation? Furthermore, is it possible to organize these differences into a typology? In fact, the "peripheral" process can be divided too into two extremes from the perspective of land systems; a) the Southeast Asian type, where people have a strong sense of personal ownership of land while communal ownership is weak, and where differentiation and disintegration of the peasant class is active, leading to a development of the process of primitive accumulation; b) the Sub-Saharan African type, where communal ownership of land is still firmly in place today, differentiation and disintegration of the peasant class is inactive, and where primitive accumulation has not progressed until very recently.

As an example of the former, I will summarize what I have already discussed on other occasions regarding trends in the Mekong Delta in Vietnam, where I conducted research. (Yamazaki 2007, 2014b, Collected Works Vol. 4) In the 1990s, there was a Leninist, classical differentiation and disintegration of the peasant class, the trends of which were determined by the disparity in rice productivity between peasants. However, since the 2000s, a new, different differentiation and disintegration of the peasant class, which could be described as a 21st century type of differentiation and disintegration, has emerged. The latter, 21st century type, is a decadent differentiation and disintegration of the peasant class that has been determined by the rampant speculative trading in the land market, which has been caused by the influx of excess capital from developed countries into Vietnam, and which has therefore become detached from the state of rice productivity. But it is also true that land speculation is giving new impetus to the process of primitive accumulation there.

The example of the Mekong Delta is by no means unique. Several previous studies have made it clear that the differentiation and disintegration of the peasant class that accompanied the "Green Revolution" is a social phenomenon that covers a wide area of Southeast Asia.[185] The differentiation and disintegration of the peasant class presupposes the existence of a market in which land is bought, sold, and rented. Moreover, such a market functions in a society where the concept of private rights to land has matured.[186]

On the other hand, when we look at the situation in sub-Saharan Africa up until recently, where peasants are protected by communities, as I have also discussed on another occasion based on on-site surveys, there has been no progress in the differentiation or disintegration of the peasant class. (Yoshida 1975, Yamazaki 2007, Collected Works Vol. 4) Therefore, there is little surplus-population pressure in the labor market there. Wages are considered relatively high compared to Southeast Asia, where the active differentiation and disintegration of the peasant class is causing surplus-population pressure in the labor market.[187]

Amin's earlier statement about the "stubborn continuity of heterogeneity in social structure" is today even stronger in the Sub-Saharan African type than in the Southeast Asian type, but what determines the difference between the

---

185 For Thailand, see Tasaka (1991). For the Philippines, see Umehara (1992). Kitahara (1985) discusses Southeast Asia in general, using Thailand and Indonesia as examples. On Indonesia, see also Geertz (1963).

186 For information on the land systems of Southeast Asian countries, see Mizuno et al. (1997).

187 Hirano (2005) emphasizes that labor costs are relatively high in Africa compared to developing Asia.

two is, on the one hand, the degree of solidity of customary communal land ownership, and, on the other, the maturity of private concepts regarding land in the consciousness of members of society, as well as the differences in the land systems that exist with the two factors.

However, if, as Amin says, the "core" process of primitive accumulation has a tendency to be unified into capital relations, then it is possible that in the "core" countries today, there is an external dependence on "peripheral" countries for labor supply, conditional on the establishment of capitalist society, i.e., the accomplishment of primitive accumulation. For the law of relative surplus-population may not be functioning well even in the "core" today. Given the enormous scale of fixed capital in capitalist enterprises, the setting of monopoly prices, and state interference in economic trends, it is not easy to believe that the problem of excess capital relative to labor can be solved by creating an excess of labor through the development of productivity and the raise of the organic composition of capital.[188] As mentioned above, from an economic perspective, primitive accumulation can be seen as a process of creating the prerequisite for the law of relative surplus-population to begin to function. However, even today, after the supply of low-wage labor from peasants to capital has dried up and this prerequisite has matured, this law seems to be prevented from fully manifesting itself due to the above-mentioned

188 The shift towards a service economy in industrial structure in developed countries in recent years is also one factor making it difficult to adopt labor-saving technologies. (Petit 1988) In addition, Shibata (1994) states that while an increase in the organic composition of capital was observed in Japan from the late 1960s to the late 1980s, the more influential factor in the decline in the rate of profit was the profit share rather than the increase in the organic composition of capital. In separate articles, I have also discussed the stagnation of Japan's capital structure (value structure) since the mid-1990s. (Yamazaki 2014a, Collected Works Vol. 2) In addition, the 2017 Economic and Fiscal White Paper (Cabinet Office 2016) points out the fact that capital equipment ratio stagnated during the same period.

technological stagnation and the shift toward a service-based economy. In addition, the law of relative surplus-population has been paralyzed by developments in social security systems such as unemployment insurance and the increased bargaining power of labor unions. (Morita 1997) As a result, capitalist companies based in the "core" seem to be forced to, on the one hand, make foreign direct investments from there to the "periphery" and become multinational corporations while hiring low-wage labor locally (in the case of Japanese companies), and, on the other hand, seek to replenish low-wage labor by introducing immigrants from the "periphery" to the "core" (in the case of Europe and the United States). [189] In other words, when considering the modern definition of "core" and "periphery," it is necessary to take into account the fact that the latter is a supplier of low-wage labor to the former.

Thus, perhaps Luxemburg's (1913) following proposition, that capitalist society needs the non-capitalist sector as a source of labor for its survival, needs to be positively reevaluated today:[190]

"Labor for this army is recruited from social reservoirs outside the dominion of capital – it is drawn into the wage proletariat only if need arises. Only the existence of non-capitalist groups and countries can guarantee such a supply of additional labor-power for capitalist production."

Incidentally, even in sub-Saharan Africa, where it seemed as though no progress was made in primitive accumulation, under communal land ownership until recent years, there have been reports that since the 2000s, as mentioned above, rapid acquisition of farmland by foreign capital and forced eviction of peasants have been taking place. In the Southeast Asian model, there was a time when the differentiation and disintegration of peasant class due to

---

189 For the immigration policies of each country, see Morita (1994) and Djajić (2001).

190 Karatani (2010) makes a similar point to that made in Luxemburg.

differences in agricultural productivity was an important aspect of primitive accumulation, but in sub-Saharan Africa, something akin to the plundering of farmland is the main aspect from the very beginning. On the other hand, the influx of excess capital from developed countries has led to land speculation within developing countries, accelerating farmers' abandonment of their farmland, as was exemplified by the Mekong Delta in Vietnam mentioned above. When we look at the current state of social differentiation and disintegration, we are tempted to wonder whether we are in a historical position where, even in developing countries, we long for a situation in which "social heterogeneity stubbornly persists" (Amin), as an idyllic world. If that is the case, then the various shades of color shown by the "peripheral" processes of primitive accumulation may be differences in the milestones toward the global end of primitive accumulation, which we can already foresee.

5. Stages of Development of World Capitalism from the Historical View from Primitive Accumulation

World capitalism is a complex mixture of national societies, each of which corresponds to one of three stages of development from the historical view from primitive accumulation, but it itself appears to have undergone the following developmental process.

First, the period from the 15th to the 18th century when primitive accumulation developed in the first-mover country, England. Let us call this the period of primitive accumulation in first-mover country.[191]

The second period is roughly from the 19th century to the first half of the

191 "Only in England has this exploitation been fundamentally accomplished." (Marx 1872-1875) "The situation in which the majority of the working population are employed is almost unique to England." (Mill 1848)

20th century, when British agriculture became capitalistic and primitive accumulation reached a certain accomplishment, while primitive accumulation also began and progressed toward accomplishment in latecomer capitalist countries other than Britain, which are today's "core." However, the primitive accumulation in the latecomer countries was not accomplished at this period. These countries experienced agricultural depressions at the end of the 19th century and after World War I, and the fall in grain prices exacerbated the economic hardship of peasants, but this also served as an opportunity for the primitive accumulation there.[192] However, the dates of primitive accumulation in these countries are only approximate and vary from country to country.[193]

192 Ouchi (1954) countered the commonly held belief among Soviet economists that the formation of a world market for agricultural products at the end of the 19th century was the cause of the agricultural depression by presenting a perspective that views the agricultural depression as part of a general depression. The formation of the world market for agricultural products occurred when grain exports from developing and newly industrialized countries began to compete with grain crops from developed countries as a result of the transportation revolution. He also explains the prolonged agricultural depression by pointing out that in addition to the fact that recessions become chronic during the imperialist stage, there are also the following special characteristics of agriculture. (a) Fixed rent determined by contract, which means it will not fall even during economic downturns. (b) The production period for biological production is long, making it difficult to adjust production flexibly. (c) The durability of family-run businesses which respond to price declines by increasing sales volume. (d) Slow technological innovation in agriculture. In addition, the theory of "the agricultural depression as part of a general depression" can also be found in Ishiwata (1953), Kurihara (1956), Miyashita (1972), and Mochida (1996). Of these, Ishiwata argues that the agricultural depression first occurred after World War I, when the capitalism of agriculture had progressed to a certain extent in major developed countries, and distinguishes it from the "agricultural depression phenomenon" that occurred at the end of the 19th century, which was caused by the influx of low-priced agricultural products into Europe.

193 In the United States, slavery was maintained in the South until the Civil War, but in other regions, capital was accumulated within the country by constantly accepting immigrants who had already been separated from the means of production, such as farmland, in the already developed countries of Europe and other regions. On the other hand, the formation of workers through domestic

For example, in France, which began capitalist development relatively early among the latecomer-developed capitalist countries, primitive accumulation through the "enclosure" of land is said to have already begun in the second half of the 18th century. (Bloch 1931) Furthermore, the movement in rural areas during the French Revolution in 1789 was a general revolt by peasants who resisted this "enclosure" and sought to restore common land.[194] In Germany, the beginning of primitive accumulation can be considered to be around the

---

primitive accumulation resulting from the differentiation of the settlers was of secondary importance compared to immigration. Furthermore, Marx (1867) described the "systematic colonization" advocated by Wakefield (1833) and actually implemented in the early days of American settlement as the "method of 'primitive accumulation' prescribed by Wakefield expressly for the use of the colonies." In "systematic colonization," 1) the government set artificially high land prices on virgin land, and 2) the money that workers paid when they purchased this land was used by the government to cover the costs of accepting immigrants from Europe and other places. It is also said that American agricultural laborers were viewed as a stepping stone to becoming small farmers. (Miyashita 1972)

194 According to Lefebvre (1939), in the second half of the 18th century, the French monarchy authorized the freedom of "enclosure" and the distribution of common land in some cantons. At this time, there was growing public interest in large-scale British-style farming. During the French Revolution, however, peasants increasingly criticized these encroachments on communal land ownership. At that time, even in France the peasantry was no longer a monolith, but compared to England the degree of differentiation and disintegration was not as advanced. Large-scale farms benefited from "enclosure" and the distribution of common land, but many peasants were critical of these. It was under these circumstances that the rural movement of 1789, which demanded the restoration of communal ownership of land and the abolition of feudal tributes, became a general peasant revolt, and which succeeded in destroying "enclosures" everywhere, restoring communal grazing, and recovering common land that had been taken from peasants. On the other hand, the following passage from *Das Kapital* shows that the increase in the number of homeless people who were cut off from the land had already become a social problem in France in the second half of the 18th century. "Even at the beginning of Louis XVI.'s reign (Ordinance of July 13th, 1777) every man in good health from 16 to 60 years of age, if without means of subsistence and not practicing a trade, is to be sent to the galleys." (Marx 1867)

time of the Stein-Hardenberg Reforms that began in 1808.[195] After that, the agricultural structure consisting of the Junkers in the East Elbe, the wealthy farmers in the West Elbe, and the part-time farmers or land-owning workers at the other extreme continued until the beginning of the 20th century.[196] On the

---

195 The following excerpts from one modern edition and one French edition of *Das Kapital* show that, due to the influence of Frederick II's reforms, primitive accumulation did not begin in Germany even at the end of the 18th century. "When Mirabeau published his book 'On the Kingdom of Prussia' (1788: quoter), serfdom still existed in most of the Prussian provinces, especially in Silesia. Nevertheless, the serfs there owned communal land." (Marx 1872-1875) "'The cleaning of estates,' or as it is called in Germany, 'Bauernlegen,' occurred in Germany especially after the 30 years' war (1618-1648: quoter), and led to peasant-revolts as late as 1790 in Kursachsen. It obtained especially in East Germany. In most of the Prussian provinces, Frederick II. for the first time secured right of property for the peasants. After the conquest of Silesia (1740: quoter) he forced the landlords to rebuild the huts, barns, etc., and to provide the peasants with cattle and implements. He wanted soldiers for his army and tax-payers for his treasury." (Marx 1867)

196 According to Kan Watanabe (1977), in 1910, the proportion of employees in agriculture, forestry, and fishing among all employees was 36%. However, in 1907, of the 9.79 million German agricultural producers, 3.49 million (36%) were employed workers, and although the remaining "own-workers" were the majority, agricultural workers included in the Junker-Landarbeiter relationship accounted for a little over one-third of the total. Moreover, in the case of peasants, the lower classes, which accounted for about three-quarters of the total, were moving completely to part-time work, and their farms were gradually being transformed into home gardens. And while keeping one foot in agriculture, peasants were transforming into proletarians. In reality, peasants were becoming more and more like workers. In addition, the number of agricultural workers was decreasing during this time due to the rapid development of German heavy industry. This led to the mechanization of farming by the Junkers and large farmers, and the shift from labor-intensive wheat production to extensive rye production. Furthermore, due to the worsening of the general "labor shortage" problem, "in the 20th century, the number of migrant workers, mainly Polish, reached approximately one million, the majority of whom were said to have been engaged in agricultural work." Despite the influx of immigrants, however, the number of agricultural workers "decreased by about 400,000 in the 25 years leading up to 1907, while agricultural wages increased by about two to three times." This note was created in order to respond to a comment by Dr. Norio Tsuge (June 2021).

other hand, in Japan, the latest among the "core" countries to develop capitalism, the end of the Edo period in the first half of the 19th century can be said to have marked the beginning of the development of primitive accumulation.[197] This second period in the primitive accumulation of world capitalism was a time of revolution and war in the latecomer developed capitalist countries, reflecting the instability of societies in the primitive accumulation stage. In particular, the two world wars in the first half of the 20th century had an external factor of conflict and strife between imperialist countries aiming to expand their economic spheres, but at the same time they were a concentrated expression of the internal instability of countries in the primitive accumulation stage.[198] Furthermore, in developed capitalist countries

---

197 Kikuchi (1994) conducted a detailed study of the course of several famines from the Tenmei famine (1782-1787) to the Tenpo famine (1833-1836). Over time, village sanctions and punishments for theft during famines became more severe, even going as far as to involve the ultimate punishment, and village rules and sanctions became more merciless and cruel. This was because frequent incidents of theft and other crimes created a sense of crisis, especially among the upper class of influential farmers, that they were threatening to destroy the existing order of the community. As a result, the upper class of farmers became overly defensive. The above describes the situation suggesting that the communal order in rural areas was in danger of collapse due to the increasing differentiation and disintegration of peasant class during this period and the movement towards privatizing common land. Furthermore, peasant uprisings and urban riots are said to have spread throughout the country after the Tenmei famine. (Kurushima 2015)

198 The global social upheaval in the first half of the 20th century was triggered by the rise of fascism within developed countries. Regarding Fascism, Gramsci (1924), who fought against Mussolini, said the following. "The middle classes, which pinned all their hopes on the fascist regime, have been crushed by the general crisis and, indeed, they themselves have become the expression of the crisis of capitalism in this period." "The distinctive feature of Fascism is its success in forming the mass organizations of the petty bourgeoisie. This is the first time in history that something like this has happened. The unique thing about Fascism is that it found a form of organization suitable for a social class that always has not been able to have a uniform composition and ideology." Furthermore, Teruoka (2003) points out the close relationship between the rise

where primitive accumulation was nearing accomplishment, as monopoly capital began to be established there, the ability to procure labor through the raising organic composition of capital within the countries was hindered, as mentioned above, and the countries became externally dependent on colonies, which were dominated by non-capitalist sectors, for labor procurement.[199] On the one hand, low-wage labor was attracted from formal and informal colonies to the colonial powers in the form of immigration, (Hilferding 1910) and on the other hand, capital was exported from developed capitalist countries to the colonies and low-wage labor was employed through the development of business activities there. (Lenin 1917) This situation aroused political passions within the developed capitalist countries, whose primitive accumulation was nearing accomplishment, for imperialist expansion outward in order to protect their own interests in the colonies. Furthermore, Hilferding (1910) testified as follows about how the development of capital relations in the colonies was linked to the use of military force in order to destroy the communities that had previously existed firmly as the traditional social foundations of the colonies and to vigorously promote primitive accumulation.

"In this way the tempo of capitalist development in new markets is greatly accelerated. The obstacle to opening up a new country is not the lack of indigenous capital, since this is eliminated by the import of capital, but in most cases quite another disruptive factor; namely, the shortage of 'free', that is to

---

of Japanese militarism and the destabilization of peasant life.

199 In England, capital exports, mainly through securities investment, began early in the 19th century, ahead of Germany and France. (Uno 1971) This early export of capital from England was used to fund the construction of transportation, communications, and public facilities in Europe, America, and the British dominions and colonies, and was also intended to enable England to import raw materials and export manufactured goods, but it can also be seen as being linked to the early accomplishment of the country's primitive accumulation. Regarding the allocation and purpose of the British securities investment, Dr. NorioTsuge provided me some comments (June 2021).

say wage, labor. The labor problem assumes an acute form, and seems to be capable of resolution only by the use of force." "As has always been the case, when capital first encounters conditions which contradict its need for valorization, and could only be overcome much too slowly and gradually by purely economic means, it has recourse to the power of the state and uses it for forcible expropriation in order to create the required free wage proletariat." "There are diverse methods of obtaining forced labor. The principal means is the expropriation of the natives, who are deprived of their land and hence of the very basis of their previous existence." "Where expropriation does not succeed immediately in such a radical way, the same end is achieved by the introduction of a system of taxation which requires the natives to make money payments on such a scale that they can only be met by incessant labor in the service of foreign capital."

After this discussion, "Finance Capital" states that "the process in which the development of capitalism freed workers for industrial employment has largely come to an end in Europe," suggesting the accomplishment of primitive accumulation in the developed countries of Europe. The ongoing capitalist expansion and the extraordinary increase in the demand for wage labor led to immigration from "south and south-east Europe and to Russia." Furthermore, "a more severe restriction of capitalist expansion in the colonial regions where there are white workers would have as a consequence that capitalism would turn increasingly to the still backward agrarian regions of Europe itself, surmounting the political barriers which stand in its way. In this way it would open up new regions where its introduction, by destroying rural domestic industry and setting free a large part of the agrarian population, would provide the material for increased emigration." From this series of quotations, it is easy to see that Hilferding has a perspective that the accomplishment of primitive accumulation in developed European countries is linked to the immigration as well as the export of capital to peripheral European countries and colonies.

In addition, Chapters 27 and 28 by Luxemburg (1913), using the British invasion of India, France's colonial policy in Algeria, and the Opium Wars as material, vividly describe the process by which the self-sufficient natural economies that had been protected by communities in the colonies were violently dismantled, and the subsequent introduction of the commodity economy transformed these formerly self-sufficient natural economies into purchasers of goods for capital and sellers of raw materials and labor.

These processes of primitive accumulation in colonies were featured by violence because they took place in a situation where there was a lack of endogenous development of agricultural productivity there, or because they were carried out "from above" by a colonial power that was parasitic on local society without waiting for that endogenous development. In other words, the use of military force was an expression of the fact that the development of primitive accumulation in the colonies was not endogenous and was therefore a bitter and difficult process involving the powerful destruction of communities.

As described above, the period marked by political instability in the latecomer developed capitalist societies and the use of violence in the colonies can be called the period of primitive accumulation in the latecomer developed capitalist countries.

The third period is from the second half of the 20th century onwards, when, while the primitive accumulation of the latecomer developed capitalist countries accomplished, the primitive accumulation of developing countries that had been liberated from their former colonial and semi-colonial status gradually began to expand along with the endogenous development of agricultural productivity triggered by events such as the Green Revolution. "Development aid" has become the watchword of the times, and while some farmers have been able to take advantage of the technical assistance provided by developed countries to developing countries, others have been left out of this trend. Economic disparities have emerged between the two groups of farmers, and differentiation and disintegration are progressing.[200] However, the validity of this explanation vary across regions within developing countries. In other words, even among developing countries, there are differences between regions where this explanation fits well and regions where it does not. In the latter half of the 20th century, Southeast Asian countries, where the idea of private ownership of land is relatively deeply rooted in society, were the center of primitive accumulation caused by competition for productivity among peasants. However, in the 21st century, as mentioned above, it appears that primitive accumulation is beginning to spread to regions of sub-Saharan Africa where communal land ownership had previously been strong. However, in the latter regions, since the foundation for the endogenous development of agricultural productivity is still lacking,[201] the violent elements of primitive accumulation come to the fore from the very start. During this period, while societies in the "core" developed countries, where the primitive accumulation have been accomplished, are relatively stable, in contrast, within developing countries, unrest and social chaos caused by revolution and war are occurring here and there, and these conditions appear to be becoming more intense.[202]

---

200 Yorozuya (2016) argues that the "resource development" system characterizes the monopoly phase that began in the early 20th century, and that the two world wars, or "total wars," can be seen as the initial stage of this.

201 "Africa rejects the Green Revolution." (Japan Association for African Studies 2014) Based on my own on-site survey, I have written papers on the productivity of native rice cultivation in the Niger Inland Delta region of Mali. (Yamazaki 2007, Collected Works Vol. 4)

202 Regarding the Iranian Revolution, a relatively recent example, Goto (2015) conducted a detailed study based on on-site research. According to the study, in Iran, during the land reform process implemented in the early 1960s, much of the land owned by landlords was divided between peasants and landlords according to product-sharing ratios. Later, mechanized agriculture developed on the farmland left to the landowners, resulting in many landless people. On the

Let us call this third period the period of primitive accumulation in developing countries. Alternatively, we could call this period the era of globalization, in the sense that capitalist production is spreading to developing countries and becoming a truly global system.

By the way, how does this three-stage division of the development process of world capitalism based on the historical view from primitive accumulation relate to the existing influential divisions of development stages of world capitalism? Furthermore, in recent years, in order to fit the economic development of East Asia in the second half of the 20th century into an analytical framework, some interesting arguments have emerged that place emphasis on the "accumulation system cycle" and "plate shifts" in understanding the history of capitalist society. Therefore, it will also be necessary to contrast the historical view from primitive accumulation with these arguments. For example, Arrighi (1994) states: Four accumulation system cycles will be discussed, each of which is characterized by a fundamental consistency in the core actors and structures of the global capital

---

other hand, although in some areas peasant proprietorship of land parcels were formed on the farmland that had been released to the peasants, in many areas pre-modern land systems such as communal ownership of farmland, open farmland systems, and land allocation systems continued. However, after the Agricultural Corporation Establishment Act was enacted in 1968, corporation farms were established on peasant-owned land in various regions, and mechanized agriculture began to be carried out there, resulting in the loss of many agricultural employment opportunities. It was in this environment that the Iranian Revolution broke out in 1979. Although the peasants initially watched the revolution taking place in the cities, they gradually became unable to remain indifferent to the landowners' properties that had been abandoned in order to escape the revolution. In addition, the landless began cultivating land owned by landlords and unoccupied land. The revolutionary government, which advocated "liberation of the oppressed," was initially ambiguous on this issue, as it was caught between its stance and Islamic law, which respects private property rights. However, the revolutionary government eventually came to allow peasants and the landless to acquire the farmland.

accumulation process. The four cycles are the Genoese cycle from the 15th to the early 17th century, the Dutch cycle from the late 16th to almost the entire 18th century, the British cycle from the late 18th to the early 20th century, and the American cycle that began in the late 19th century and has continued through to the current period of monetary expansion. Arrighi further states that each cycle contains a period of production expansion (the MC phase of capital accumulation) and a period of financial recovery and expansion (the CM phase). Arrighi also sees in the current situation a crisis for the American system and the possibility that East Asia may become the center of the global market.[203] Next, Obata (2011) advocates the "multiple origins theory," stating that capitalist societies "emerge in clusters in different countries and regions at different times, and serve as a catalyst for development into a new stage," and describes the transition from the mercantilist stage on the continent, based on the wool industry, to the liberal stage in England, based on the cotton industry, as a "plate change." As an extension of this argument, recent globalization (including the aspect that a new movement towards capitalist society has progressed rapidly outside of developed capitalist countries) can also be seen as a "new plate change" (the large plate of the 20th century that carried the Cold War structure is being replaced by a new plate). Compared to the perspective advocated by Arrighi and Obata that emphasizes "cycle" and "plate change," the historical view from primitive accumulation places a much greater emphasis on linear development. However, this view shifts the focus away from capital, which has been emphasized in traditional theories of development stages, and towards the opposite pole, labor supply.

In Uno's theory (1971), the world-historical stages of development of capitalist society are divided into the mercantilist stage (corresponding to early commercial capital and early usury capital), the liberal stage (industrial

203 A similar circular understanding is taken by Nitta (2018).

capital), and the imperialist stage (finance capital), based on the dominant form of capital and the corresponding economic policies. Furthermore, while viewing the period after World War I as already being a transitional period towards a socialist society in world history, Uno treats it as the subject of his "analysis of the current situation" as "capitalism against socialism."

The period of primitive accumulation in first-mover country of the historical view from primitive accumulation, which focuses on the relationship between the capitalist sector and non-capitalist areas, can be said to correspond to Uno's the mercantilist stage. Furthermore, the period of primitive accumulation in the latecomer countries of the historical view from primitive accumulation roughly corresponds to the liberal and imperialist stages of Uno's theory, as well as the period of analysis of the current situation up until World War II. Furthermore, the period of primitive accumulation in developing countries in the historical view from primitive accumulation corresponds to the period of analysis of the current situation in Uno's theory, particularly the period after World War II. The historical view from primitive accumulation places more importance on World War II as a turning point than World War I, because it places more importance on trends in developing countries than on the socialist countries of Europe.

Furthermore, in Uno's theory of economic principles, he sees the fundamental impossibility of capitalist society in the commodification of labor-power, but in his classification of the stages of development of capitalist society, he bases it on the dominant capital and the corresponding form of economic policy, and here it seems that there is a methodological gap between his theory of economic principles and his theory of stages of development. Indeed, Uno (1971) describes the economic policies of the mercantilist stage as promoting the process of separation between direct producers and the means of production and the resulting process of forging the propertyless as laborers, emphasizing the aspect of "enforcing a primitive commodification of labor-power." Uno also states that during the liberal stage, capital "was able to secure the commodification of labor-power by its own efforts as the reproduction process developed," suggesting the background to why economic policies became "passive" at this stage. However, Uno argues that at the imperialist stage, the organic composition of capital constantly became more increased, resulting in a chronic surplus of the labor force, but the relationship between these characteristics of this stage of the commodification of labor-power and the capital export policy that characterized this stage does not seem to be entirely clear.

On the other hand, like Polanyi (1944) and Uno's theory of economic principles, the historical view from primitive accumulation also seeks to see the fundamental impossibility of capitalist society in the commodification of labor-power, but at the same time attempts to make this perspective consistent even in its recognition of the stages of development of capitalist society by emphasizing the source of labor-power supply. Furthermore, if we try to view the history of capitalist society as a process of purifying and impurifying capital relations, as in Uno's theory, there seems to be a problem in that we cannot grasp the collapse of capitalist society. However, in the historical view from primitive accumulation, we can simply consider the end of global primitive accumulation to be the end of capitalist society. This is because modern capitalist society, which is dependent on the outside world in terms of labor and in that sense is not independent, will no longer have an "outside" to rely on with the end of global primitive accumulation.

Bibliography

Adamczewski A., Hertzog T., Dosso M., Jouve P. et Jamin J.Y. (2011) L'irrigation peut-elle se substituer aux cultures de décrue ?, *Cahiers Agricultures*, 20 (1-2), pp. 97-104, doi: 10. 1684/agr. 2011. 0469.

Amin S. (1970) *L'accumulation à l'échelle mondiale*, Editions Anthropos, Paris.

Amin S. (1972) Le modèle théorique d'accumulation et de développement dans le monde contemporain: La problématique de transiotion, *Tiers-Monde*, 13 (52), pp. 703-726.

Amin S. (1973) *Le développement inégal: Essai sur les formations sociales du capitalisme périphérique*, Éditions de Minuit, Paris.

Amin S., Arrighi G., Frank A.G. et Wallerstein I. (1982) *La crise, quelle crise ? Dynamique de la crise mondiale*, Éditions Maspéro, Paris.

Aristotle (edited around 300 BC) *Nicomachean ethics*.

Arrighi G. (1994) *The Long Twentieth Century: Money, Power and the Origins of Our Times*, Verso, New Tork.

Barbier B., Ouedraogo H., Dembélé Y., Yacouba H., Barry B. et Jamin J.Y. (2011) L'agriculture irriguée dans le Sahel Ouest-Africain, *Cahiers Agricultures*, 20 (1-2), pp. 24-33, doi: 10. 1684/agr. 2011. 0475.

Barrière O. et Barrière C. (2002) *Un droit à inventer: Foncier et environnement dans le Delta Intérieur du Niger (Mali)*, IRD Editions, Paris.

Bauer O. (1907) *Die Nationalitätenfrage und die Sozialdemokratie*, Verlag der Wiener Volksbuchhandlung Ignaz Brand, Wien.

Bauer O. (1913) *Die Akkumulation des Kapitals*.

Bélières J-F., Hilhorst T., Kébé D., Keïta M.S., Keïta S. et Sanogo O. (2011) Irrigation et pauvreté: Le cas de l'Office du Niger au Mali, *Cahiers Agricultures*, 20 (1-2), pp. 144-149, doi: 10. 1684/agr. 2011. 0473.

Bible.

Bloch M. (1931) *Les caractères originaux de l'histoire française*, H. Ashehoug & co., Oslo.

Böhm - Bawerk E. (1896) *Zum Abschluss des Marxschen Systems*, O. Haering, Berlin.

Bonneval P., Kuper M. et Tonneau J.P. (2002) *L'Office du Niger: Grenier à riz du Mali*, CIRAD/Karthala, Montpellier.

Bradby B. (1975) The destruction of natural economy, *Economy and Society*, 4 (2), pp. 127-161.

Broadberry S., Campbell B. M. S., Klein A., Overton M. and Leeuwen B. v. (2015) *British economic growth 1270-1870*, Cambridge University Press.

Brocheux P. (1995) *The Mekong Delta: Ecology, economy and revolution, 1860-1960*, Center for Southeast Asian Studies, University of Wisconsin-Madison.

Brondeau F. (2011) L'agrobusiness à l'assaut des terres irriguées de l'Office du Niger (Mali), *Cahiers Agricultures*, 20 (1-2), pp. 136-43, doi: 10. 1684/agr. 2011. 0472.

Caesar J. (52-51 BC) *Commentaries on the Gallic war.* (Translated in English by W.A. McDevitte and W.S. Bohn)

Cho K. and Yagi H. (eds.) (2001) *Vietnamese agriculture under market-oriented economy*, Agricultural Publishing House, Hanoi.

Cissé S. (1982) Les leyde du Delta Central du Niger: Tenure traditionnelle ou exemple d'un amenagement de territoire classique, In Bris E.L., le Roy E. et Leimdorfer F. (éds.) *Enjeux fonciers en Afrique Noire*, Karthala, Paris, pp.178–189.

Cormier-Salem M-C. (éd.) (1999) *Rivières du Sud: Sociétés et mangroves Ouest-Africaines*, IRD, Paris.

Coulibaly Y., Bélières J-F. et Koné Y. (2006) Les exploitations agricoles familiales du grand périmètre irrigué de l'Office du Niger au Mali: Evolutions et perspectives, *Cahiers Agricultures*, 15 (6), pp. 562-569.

Cunow H. (1926-1931) *Complete history of economy.*

De Noray M.L. (2003) Delta Intérieur du Fleuve Niger au Mali – Quand la crue fait la loi : L'organisation humaine et le partage des ressources dans une zone inondable à fort contraste, *VertigO - la revue électronique en sciences de l'environnement* [En ligne], 4 (3), http://vertigo.revues.org/3796.

Devereux S. (1993) *Theories of famine*, Prentice-Hall Europe.

DIRASSET (1994) *Avant-projet de schémas régionaux d'aménagement et de développement: District de Mopti*, Mopti (Mali) .

Djajić S. (ed.) (2001) *International migration*, Routledge, London,.

Dobb M. (1946) *Studies in the development of capitalism*, Routledge & Kegan Paul, London.

Dobb M., Hill C., Hobsbawm E., Lefebvre G., Merrington J., Procacci G., Sweezy P., Takahishi K. and Hilton R. (1976) *The transition from feudalism to capitalism,* New Left Books, London.

Eckstein G. (1913) *Die Akkumulation des Kapitals*, Vorwaris.

Emmanuel A. (1969) *L'échange inégal: Essai sur les antagonismes dans les rapports économiques internationaux*, François Maspero, Paris.

Engels F. (1853) *Letter to Marx dated June 6, 1853.*

Engels F. (1875) *The social condition of Russia.*

Engels F. (1878) *Herrn Eugen Dührings Umwälzung der Wissenschaft.*

Engels F. (1884) *Letter to Kautsky dated February 16, 1884.*

Febvre L. (1922) *La terre et l'évolution humaine*, La Renaissance du Livre, 1922.

Fröbel F., Heinriches J. and Kreye O. (1980) *The new international division of labor: Structural unemployment in industrialized countries and industrialization in developing countries,* Cambridge University Press.

Gafsi M., Dugué P., Jamin J.Y. et Brossier J. (éds.) (2007) *Exploitations agricoles familiales en Afrique de l'Ouest et du Centre: Enjeux, caractéristiques et éléments de gestion*, Editions Quae, Versailles.

Gallais J. (1984) *Hommes du Sahel: Espaces-temps et pouvoirs: Le Delta Intérieur du Niger, 1960-1980*, Flammarion, Paris.

Geertz C. (1963) *Agricultural involution: The processes of ecological change in Indonesia*, University of California Press.
Gramsci A. (1924) Crisi in Italia, *Ordine Nuovo*.
Griaule M. (1948) *Dieu d'eau; Entretiens avec Ogotemmêli*, Fayard, Paris.
Grossmann H. (1929) *Das Akkumulations- und Zusammenbruchsgesetz des kapitalistischen Systems*, Verlag von C. L. Hirschfeld, Leipzig.
Hilferding R. (1910) *Das Finanzkapital. Eine Studie über die jüngste Entwicklung des Kapitalismus*, Wiener Volksbuchhandlung, Vienna. (Translated in English by Tom Bottomore)
Hobsbawm E. J. E. (ed.) (1964) *Karl Marx: Pre-capitalist economic formations*, Lawrence & Wishart.
Huberman L. (1936) *Man's worldly goods*, Harper & Brothers.
Hudson P. (1992) *The Industrial Revolution*, Edward Arnold (Publishers) Limited.
Hugo V. M. (1830) *Hernani*.
Hugo V. M. (1874) *Quatrevingt-treize*.
Hunt E. H. (1986) Industrialization and regional inequality: Wages in Britain, 1760-1914, *Journal of Economic History*, 46 (4), pp. 935-966.
Hyden G. (1986) The anomaly of the African peasantry, *Development and Change*, 17, pp. 677-705.
Iliffe J. (1983) *The emergence of African capitalism*, Macmillan Press Limited, London.
Institute of Economics of the Academy of Sciences of the Soviet Union (1954) *Textbook of economics*.
Jamin J.Y., Bouarfa S., Poussin J.C. et Garin P. (2011) Les agricultures irriguées face à de nouveaux défis, *Cahiers Agricultures*, 20 (1-2), pp. 10-15, doi: 10. 1684/agr. 2011. 0477.
Keller C. (1919) *Die Stammesgeschichte unserer Haustiere*, 2 Auflage.
Kuper M. (2011) Des destins croisés: Regards sur 30 ans de recherches en grande hydraulique, *Cahiers Agricultures*, 20 (1-2), pp.16-23, doi: 10. 1684/agr. 2011. 0467.
L'Afrique Authentique (2003) *Le petit futé du Mali: édition 2003*, Lyon.
Lefebvre G. (1939) *Quatre-vingt-neuf*, Editions Sociales.
Lenin V. (written in 1893) *Regarding so-called market issues*.
Lenin V. (1917) *Imperialism, the highest stage of capitalism*.
Lévi-Strauss C. (1955) *Tristes tropiques*, Librairie Plon, Paris.
Lévy-Bruhl L. (1910) *Les fonctions mentales dans les sociétés inférieures*, Félix Alcan, Paris
Lewis W. A. (1954) Economic development with unlimited supplies of labor, *The Manchester School*, 22 (2), pp.139-191.
Locke J. (1690) *Two treaties of government*.
Luxemburg R. (1899) *Sozialreform oder Revolution?*
Luxemburg R. (1913) *Die Akkumulation des Kapitals,—Ein Beitrag zur Ökonomischen Erklärung des Imperialismus*, Frankes Verlag. (Translated in English by Agnes Schwarzschild)

Luxemburg R. (1921) *Die Akkumulation des Kapitals, oder Was die Epigonen aus der Marxschen Theorie gemacht haben,* Frankes Verlag .
Magasa A. (1978) *Papa-commandant a jeté un grand filet devant nous: L'Office du Niger 1902-1962,* François Maspero, Paris.
Marx K. (1853) The British rule in India, the *New-York Daily Tribune, June 25,* 1853.
Marx K. (written in 1858) *Formen die der Kapitalistischen Produktion vorhergehen.* (Translated in English by Cohen J.)
Marx K. (written in 1862-1863) *Theorien über den Mehrwert.*
Marx K. (1867) *Das Kapital Buch1,* Verlag von Otto Meisner. (Translated in English by Moore S. and Aveling E. and edited by Engels F.) (Moscow version)
Marx K. (1872-1875) *Le Capital,* Maurice Lachatre.
Marx K. (written in 1881) *Reply to V. I, Zasulich's letter.*
Marx K. (1885) *Das Kapital Buch 2.* (Moscow version)
Marx K. (1894) *Das Kapital Buch 3.* (Moscow version)
Maswood S. J. (2018) *Revising globalization and the rise of global production networks,* Palgrave Macmillan.
Mauss M. (1925) *Essais sur le don: Forme et raison de l'échange dans les sociétés archaïques* , L›Année Sociologique, Paris.
Meillassoux C. (1960) Essai d'interprétation du phénomène économique dans les sociétés traditionnelles d'auto-subsistance, *Cahiers d'Etudes Africaines,* 1 (4), pp.38-67.
Meillassoux C. (1964) *Anthropologie économique des Gouro de Côte d'Ivoire: De l'économie de subsistance à l'agriculture commercial,* Ecole Pratique des Hautes Etudes, Paris.
Meillassoux C. (1975) *Femmes, greniers et capitaux,* Maspero, Paris.
Mies M., Benholdt-Thomsen V. and Werlhof C. v. (1995) *Women: The last colony,* Zed Books Ltd.
Mill J. S. (1848) *Principles of political economy,* John W. Parker.
Montesquieu C.-L. (1748) *De l'esprit des lois,* Barrillot & Fils, Genève.
More T. (1516) *Utopia.*
Morgan L. H. (1881) *Houses and house-life of the American aborigines.*
Ominami C. (1986) *Le tiers monde dans la crise,* La Découverte, París.
Petit P. (1988) *La croissance tertiaire,* Economica, Paris.
Polanyi K. (1944) *The great transformation,* Beacon Press.
Polanyi K. (1957) The economy as instituted process, in Polanyi K., Arensberg C. M. and Pearson H. W. (eds.) *Trade and market in the early empires,* The Free Press, Glencoe, pp. 243-270.
Portères R. (1950) Vieilles agricultures de l'Afrique intertropicale: Centres d'origine et de diversification variétale primaire et berceaux d'agricultures antérieures au X Ⅵe siècle, *Agronomie Tropicale,* 10, pp. 489-507.
Quran.
Ricardo D. (1817) *On the principles of political economy and taxation.*
Rosenberg (1931-1933) *Commentary on Capital.*

Sahlins M. (1972) *Stone age economics*, Aldine Publishing Co.
Schmitz J. (1986) L'Etat géomètre: Les leydi des Peuls du Fuuta Tooro (Sénégal) et du Massina (Mali), *Cahiers d'Etudes Africaines*, 26 (3), pp.349-394.
Scott J. C. (1976) *The moral economy of peasant: Rebellion and subsistence in Southeast Asia*, Yale University.
Smith A. (1759) *The theory of moral sentiments*, printed for Andrew Millar, in the Strand; and Alexander Kincaid and J. Bell, Edinburgh .
Smith A. (1776) *An inquiry into the nature and causes of the wealth of nations*, W. Strahan and T. Cadell, London.
Suttanipāda
Tacitus P. C. (AD 98) *Germania*. (Translated in English by Herbert W. Benario)
Tolokonski and Kraus (1930) *Theory of capital accumulation and depression*.
Vidal de la Blache P. (1922) *Principles de géographie humaine*, Armand Colin, Paris.
Wakefield E. G. (1833) *England and America, A comparison of the social and political state of both nations*, London.
Wallerstein I. (1974) *The modern world-system: Capitalist agriculture and the origins of the European world-economy in the sixteen century*, Acacemic Press.
Weber M. (1920) *Das antike Judentum*.
Weber M. (1920-1921) *Gesammelte Aufsätze zur Religionssoziologie*.
Weber M. (1924) *Abriß der universalen Sozial- und Wirtschaftsgeschichte, Aus den nachgelassenen Vorlesungen herausgegebon von Prof. S. Hellmann und Dr.M. Palyi*, 2te Auflage, München und Leipzig.
Wittfogel K. A. (1957) *Oriental despotism: A comparative study of total power*, Yale University Press.
World Bank (1993) *East Asia miracle: Economic growth and public policy*.
Yamazaki R. (2004) *Agriculture in the Mekong Delta of Vietnam*, Louma Productions, Aniane (France) . [Collected Works Vol. 3 (2022) includes Japanese translations of Chapters 1, 2, and 6 of the book.]

Japanese text

Akabane H. (1971) *Introduction to the analysis of underdeveloped economics*, Iwanami Shoten.
Akita S. (2012) *History of the British Empire*, Chukoshinsho.
Arai H. (1985) Debate regarding the establishment and significance of the theory of relative surplus-population, Tomizuka R., Fumio H. and Honma Y (eds.), *System of Capital 3: Surplus-value and capital accumulation*, Yuhikaku, pp. 443-466.
Aramata S. (1972) *Law of value and wage labor: Introduction to research on wage labor theory*, Kouseisha Kouseikaku.
Baba K. (1981) *Perspective on modern capitalism*, Tokyo University Press.
Baba, K. (1986) *Enrichment and financial capital*, Minerva Shobo.
Cabinet Office (ed.) (2016) *2017 Economic and fiscal white paper*, Nikkei Printing.
Cho K., Senda N., Hirata S., Egashira T., Murata A. and Kato K. (1979) Agriculture and water use development in Mali, *Journal of the Agricultural Engineering Society*,

Japan, 47 (11), pp. 51-55.

Chuo University/Japan Institute of Comparative Law (1978) *Invitation to Islamic law: Record of the Islamic law lecture (Tokyo) : July 4-7 1977*, Chuo University Press.

Fujise K. (1985) Types of primitive accumulation, Tomizuka R., Fumio H. and Honma Y (eds.), *System of Capital 3: Surplus-value and capital accumulation*, Yuhikaku,, pp. 293-308.

Fujita M. (2021) Net capitalism and prospects for social change, *Political Economy Quarterly*, 58 (2), pp. 18-33.

Fukumoto K. (2003) Japan's controversy of Asiatic mode of production, Second controversy edition: 1965-1982, *Meiji University Liberal Arts Collection*, 367.

Fukutomi M. (ed.) (1969) *Revival of the Asian production mode controversy*, Miraisha.

Fukutomi M. (1970) *Community controversy and the principle of property*, Miraisha.

Hamamura K., Kitamura Y. and Sawada H. (1992) *Survey report on the characteristics of agriculture and forestry in West Africa: Niger, Mali, and Cote d'Ivoire*, Tropical Agriculture Research Center, Ministry of Agriculture, Forestry and Fisheries.

Hazama M. (1953) Determination of grain prices in a peasant proprietorship of land: Is peasant labor social labor? Collected papers to commemorate Professor Hyoe Ouchi's retirement, *Research on Marxian economics (Part 1)*, Iwanami Shoten.

Hidaka S. (1983) Economic Principles, Yuhikaku.

Hirano K. (2005) Poverty link between agriculture and industry, Hirano K. (ed.), *Empirical analysis of the African economy*, Institute of Developing Economies, pp. 131-190.

Hirano K. (2013) *Economic continent Africa: From resource and food issues to development policy*, Chuko Shinsho.

Hirose S. and Wakatsuki T. (eds.) (1997) *Restoration of the ecological environment of the West African savannah and revitalization of rural areas*, Agricultural and Forestry Statistics Association.

Horie E. (ed.) (1962) *Studies on the English Revolution*, Aoki Shoten.

Ichihara K. (1990) Development of reproduction theory after Marx, Tomizuka R. and Imura K. (eds.), *System of Capital 4: Circulation and reproduction of capital*, Yuhikaku, pp. 444-482.

Iinuma J. (1970) *Fudo and history*, Iwanami Shoten.

Inoue M. (1979) Commentary, Watsuji T., *Fudo: Anthropological consideration*, Iwanami Shoten, pp. 289-298.

Inuzuka S. (ed.) (1982) *Late Showa period agricultural issues collection 11: Agricultural price theory*, Nobunkyo.

Ishimoda T. (1946) *Formation of the medieval world*, Iwanami Bunko.

Ishiwata S. (1953) *Theory of agricultural depression*, Rironsha.

Ishiwata S. (1959) Land ownership and agricultural product price formation in modern times: Land ownership and agricultural product price formation after agrarian reform, published in commemoration of Dr. Yasuo Kondo's 60th birthday, *Study on the theory of rent in Japanese agriculture*, Yokendo.

Ito M. (1982) *Modern Marxian economics*, TBS Britannica.
Ito, M. (1990) *Capitalism in reverse flow*, Toyo Keizai Shinposha.
Ito M. (2018) *Introduction to capitalist economy*, Heibonsha
Iwasaki T. (2015) Book review: Ryoichi Yamazaki's "Agricultural structural dynamics under globalization: Various types of primitive accumulation," *Journal of Rural Economics*, 87 (3), pp. 314-316.
Iwata H. (1964) *World capitalism*, Miraisha.
Izutsu T. (1958) *Commentary, Quran*, Iwanami Bunko.
Izutsu T. (1981) *Islamic culture: What lies at its root*, Iwanami Bunko.
Japan Association for African Studies (ed.) (2014) Encyclopedia of African studies, Showado.
Japan Association for International Collaboration of Agriculture and Forestry (1986) *Agriculture in Mali: Current status and development challenges*, Overseas Agricultural Development Research Country Study Series, 26.
Karatani K. (2010) *Structure of world history*, Iwanami Shoten.
Kawakami H. (1929) *Introduction to Capital*, Aoki Bunko.
Kawata J. (1981) *Savannah notebook*, Shincho Sensho.
Kawata J. (ed.) (1997) *Nature and culture of the Great Bend of the Niger River*, University of Tokyo Press.
Kawata J. (ed.) (1999) *Introduction to Africa*, Shinshokan.
Kikuchi I. (1994) *Social history of famine*, Azekura Shobo.
Kinoshita E. (ed.) (1960) *Controversy and theory of international values*, Kobundo.
Kitahara J. (1985) *Development and agriculture: Capitalism in Southeast Asia*, Sekai Shisosha.
Kobayashi N. (1977) *Establishment of the Wealth of Nations*, Miraisya.
Koga M. (1977) Relative surplus-population or industrial reserve army, in Sato K., Okazaki E., Furihata S. and Yamaguchi S. (eds.), *Learning about Capital II*, Yuhikaku, pp. 212-227.
Kotani H. (1979) *Marx and Asia*, Aoki Shoten.
Kotani H. (1982) *Community and modernity*, Aoki Shoten.
Kumashiro Y. (1969) *Comparative study of agricultural technology: Traditional farming in East Asia and modern farming in Western Europe*, Ochanomizu Shobo.
Kurihara H. (1955) Agricultural product price policy and production costs, *Agricultural statistics research materials 17*, Statistics Research Group.
Kurihara H. (1956) *On the agricultural depression*, Aoki Shoten.
Kurushima H. (2015) Peasant uprisings and urban riots, in Otsu T., Sakurai E., Fujii J., Yoshida H., and Li S. (eds) *Iwanami Lectures on Japanese History, Volume 13, Early Modern Period 4*, Iwanami Shoten, pp. 209-250.
Matsumoto H. (1995), *International value theory and floating exchange rates*, Shinsensha.
Matsuo T. (1978) *Pre-capitalist production mode theory*, Ronsosha.
Mishima T. (2005) *Inheritance of agricultural market theory*, Nihon Keizai Hyoronsha.
Miyamoto S. and Matsuda M. (eds.) (1997) *Shinsho African history*, Kodansha.
Miyashita S. (1972) *Capitalism and agricultural depression*, Hosei University Press.

Miyazaki Y. (1697) *Encyclopedia of agriculture,* Iwanami Bunko.
Miyazawa K. (2018) Capital accumulation and long-term changes in industrial reserve army, *Political Economy Quarterly*, 55 (2), pp. 41-52.
Mizuno H. and Shigetomi S. (eds.) (1997) *Economic development and land systems in Southeast Asia*, Institute of Developing Economies.
Mizuta H. (2001) Explanation, Smith A., *The wealth of nations*, Iwanami Bunko
Mochida K. (1996), *World economy and agricultural issues*, Hakuto Shobo.
Mochizuki S. (1977) Primitive accumulation: Criticism of the bourgeois defense of property, Sato K., Okazaki E., Furihata S. and Yamaguchi S., *Learning about Capital II Volume 1: The production process of capital* (*Part 2*), Yuhikaku, pp.270-286.
Mochizuki S. (1981a) Theory of production-mode articulation: A key to the history and modernity of the third world, *Keizai Hyoron*, 30 (7), pp. 104-119.
Mochizuki S. (1981b) Third world studies and the theory of primitive accumulation, *Keizai Hyoron,* 30 (12), pp. 86-101.
Mochizuki S. (1982) Perspective and axis of primitive accumulation theory, *Shiso*, 695, pp. 79-95.
Mochizuki S. (2000) Pre-capitalist economic formations in the Grundrisse, Hattori F. and Sato K. (eds.), *System of Capital 1: Establishment of the System of Capital*, Yuhikaku, pp. 293-305.
Mori O. (1915) *Sansho Daiyu.*
Morita K. (ed.) (1994) *International labor mobility and foreign workers*, Dobunkan.
Morita K. (1997) *Composition of world economics*, Yuhikaku.
Motoyama Y. (1976) *World economic theory*, Dobunkan.
Motoyama Y. (1982) *Introduction to trade theory*, Yuhikaku.
Muraoka S. (1976) *Marx's theory of the world market*, Shin Hyoron.
Muraoka S. (1988) *World economic theory*, Yuhikaku.
Muroi Y. (1995) Overview: Structure and transformation of the world economy, Morita K (ed.), *World economic theory*, Minerva Shobo, pp. 1-51.
Nakao S. (1966) *Origin of cultivated plants and agriculture*, Iwanami Shoten.
Nakao S. (1969) *A journey in search of the origins of agriculture: From the Niger to the Nile*, Kodansha.
Nakata K. (1923) *Private law as seen in the literature of the Tokugawa period*, Iwanami Shoten.
Naruse N. (1985) Controversy over the international theory of value, Kinoshita E. and Muraoka S. (eds.), *System of Capital 8: States, international commerce, and world markets*, Yuhikaku, pp. 365-385.
Nawa T. (1949) *Research on international values*, Nippon Hyoronsha.
Nitta S. (2018) Long cycle of world capitalism and recursion of principle compatible / incompatible situations, *Political Economy Quarterly*, 55 (1), pp. 15-24.
Noro E. (1930) *History of the development of Japanese capitalism*, Iwanami Shoten
Obata M. (2011) Fundamental problems in Marxian economic theory under the rise

of the emerging capitalism, *Political Economy Quarterly*, 48 (1), pp. 15-25.
Ogawa R. (1987) *Living in the Sahel: West African Fulbe ethnography*, NHK Books.
Oji T. (1993) Rice cultivation in the Niger River delta: From the perspective of an Asian researcher, in Sasaki T. (ed.), *Agricultural technology and culture*, Shueisha, pp. 66-81.
Okishio N. (1980) *Issues in the analysis of modern capitalism*, Iwanami Shoten.
Onozuka T. and Numajiri A. (eds.) (2007) *Rereading Hisao Otsuka's "Basic theory of community,"* Nihon Keizai Hyoronsha.
Otani T. and Tairako T. (2013) *Reading Marx from excerpted notes of Marx: Editing of MEGA Division IV and studying included notes*, Sakurai Shoten.
Otsuka H. (1955) *Basic theory of community*, Iwanami Shoten.
Otsuka H. (1956a) *Economic history of Europe*, Kobundo.
Otsuka H. (1956b) How to problematize the community, *Sekai*, March/April issue.
Otsuka H. (1960) Introduction, Otsuka H., Takahashi K. and Matsuda T. (eds.) *Lecture on Western economic history I: Economic foundations of feudalism*, Iwanami Shoten
Otsuka H. (1962) Basic conditions for the dissolution of community: A theoretical consideration, Dr. Yoshio Motoida's Koki Memorial Collected Papers Publication Committee (ed.), *Research on Western economic history and intellectual history*, Sobunsha.
Ouchi T. (1954) *Agricultural depression*, Yuhikaku.
Ouchi T. (1977) *Agricultural economics*, Chikuma Shobo.
Sakurai E. (2011) *Economics of gifts: Between ritual and economy*, Chuko Shinsho.
Sakurai T. (2005) Land ownership system of lowland wetlands and investment in water management technology in West Africa, *Journal of Rural Economics*, 76 (4), pp. 241-250.
Sakurai T. (2009) *Agricultural origins of capitalism and economics*, Syakai Hyoronsha.
Sasaki T. (1993) Issues and methods of modern world economic theory, in Muraoka S. and Sasaki T. (eds.) *Structural change and the world economy*, Fujiwara Shoten.
Sasaki Y. (ed.) (2000) *New edition: Introduction to livestock science*, Yokendo.
Sato H. (1983-1984) The logic of rent in the world market, *Artes liberales*, 33, pp. 145-172; 34, pp. 95-117.
Sato H. (1994) *Basic logic of foreign trade*, Sofusha.
Sato K., Okazaki E., Furihata S. and Yamaguchi S. (eds.) (1977) *Learning about Capital* I-V, Yuhikaku.
Shibata T. (1994) Determinants of fluctuations in the rate of profit, *Economic Sciences*, 41 (2), pp. 21-32.
Shiina S. (1962) *Agricultural structure during the British Industrial Revolution*, National Research Institute of Agricultural Economics, Research series 67.
Shiina S. (1976) *Agricultural thought: Marx and Liebig*, Tokyo University Press.
Shimada S. (1975) Cultivation forms and land tenure in Nigeria: Focusing on Eastern Nigeria, Yoshida M. (ed.) *Agriculture and land tenure in Africa*, Institute of Developing Economies, pp. 85-124.

Shimizu M. (1942) *Japanese medieval villages*, Iwanami Bunko.
Shinozuka S. (1974) *Land ownership and modern times: Perspectives from history*, NHK Books.
Sugimura K. (2004) Economics of African peasants: Regional comparison of organizational principles, Sekai Shisosha.
Taguchi F. (1979) *New developments in Marxist theory of the state*, Aoki Gendai Series.
Takahashi A. (1999) Translator's afterword, in Scott J. C., *The moral economy of peasant: Rebellion and subsistence in Southeast Asia*, Keiso Shobo, pp. 299-306.
Takeuchi S. (ed.) (2017) *Land and power in contemporary Africa*, Institute of Developing Economies.
Takezawa S. (1984) African Rice, *Anthropology Quarterly*, 15 (1), pp. 66-116.
Takezawa S. (1988) "Water Spirits" and Islam: Social and religious changes among the Bozo people, *National Museum of Ethnology Research Report*, 13 (4), pp. 857-896.
Takezawa S. (2008) *Savannah river people: Ethnography of memory and narration*, Sekai Shisosha.
Takezawa S. (ed.) (2015) *58 Chapters for understanding Mali*, Akashi Shoten.
Takezawa S., Cissé M. and Oda H. (2005) Mema in the history of West African: Economic bases of ancient Ghana and Mali, *Journal of African Studies*, 66, pp. 31-46.
Tamagaki Y. (1969) A Study on Marx's theory of accumulation, *Senshu Keizai Ronshu*, 7, pp. 1-56.
Tamanoi Y. (1978) *Economy and ecology: The road to economics in a broad sense*, Misuzu Shobo.
Tasaka T. (1991) *Research on the differentiation of Thai farmers*, Ochanomizu Shobo.
Tashiro T. (1968) Rent for peasants, Tsuru D. et al. (eds.), *Economic development and peasant laws III*, Ochanomizu Shobo.
Teruoka S. (ed.) (2003) *150 years of Japanese agriculture*, Yuhikaku.
Tomizuka R., Hattori F. and Honma Y. (eds.) (1985), *System of Capital 3: Surplus-value and capital accumulation,* Yuhikaku.
Tomizuka R. (1990) Structure and dynamics of extended reproduction [1], Tomizuka R. and Imura K. (eds.), *System of Capital 4: Circulation and reproduction of capital*, Yuhikaku, pp. 271-285.
Tsuge N. (2010) *Symbiotic agricultural system in western capitalist countries: Relationship between symbiotic principles and agriculture from a focus on the UK,* Association of Agriculture and Forestry Statistics.
Uchida Y. (1961) *Lecture on the history of economics*, Miraisya.
Uchida Y. (1996) Translator's afterword: About Weber's Anscient Judaism, in Weber M., *Ancient Judaism*, Iwanami Bunko, pp. 1013-1106.
Umehara H. (1992) *Rural areas in the Philippines*, Kokon Shoin.
Umesao T. (1967) *Ecological historical view of civilization*, Chuokoron Shinsha.
Uno K. (1932) Basic considerations on the reproduction schema, *Uno Kozo Collected Works Vol. 3*, Iwanami Shoten, pp. 102-127.

Uno K. (1936) *Economic policy theory* (*Part 1*), Kobundo.
Uno K. (1953) *Depression*, Iwanami Shoten.
Uno K. (1962) *Economic methodology*, Tokyo University Press.
Uno K. (ed.) (1967) *Research on Capital II: Surplus-value and accumulation*, Chikuma Shobo.
Uno K. (1971) *Economic policy theory, Revised edition*, Kobundo.
Watanabe K. (1977) Development process of German agriculture, in Ouchi T. (ed.), *Agricultural economics*, Chikuma Shobo, pp.117-193.
Watanabe T. (1977) *Rice road*, NHK Books.
Watanabe T. (1996) *Development economics 2nd edition*, Nippon Hyoronsha.
Watsuji T. (1935) *Fudo: Anthropological consideration*, Iwanami Shoten.
Yamada M. (1934) *Analysis of Japanese capitalism*, Iwanami Syoten.
Yamada M. (1951) Introduction to analysis of the reproduction schema, *Yamada Moritaro Collected Works Volume 1*, Iwanami Shoten, pp. 53-274.
Yamada M. (1962) *Fundamental analysis of the reproduction structure of Japanese agriculture*, Land System Archives.
Yamada T. (2008) *Various capitalisms: Comparative capitalism analysis*, Fujiwara Shoten.
Yamazaki R. (1996) *Regional characteristics of the labor market and agricultural structure*, Agriculture and Forestry Statistics Association. [Collected Works Vol. 1 (2020) Part 1]
Yamazaki R. (2007) *Symbiotic agricultural systems in peripheral developing countries: Focusing on Southeast Asia and Africa*, Agricultural and Forestry Statistics Association. [Collected Works Vol. 4 (2021) Part 1]
Yamazaki R. (2014a) Capital accumulation and agriculture in the postwar Japanese economy, in R. Yamazaki, *Agricultural structural dynamics under globalization*, Ochanomizu Shobo, pp.145-197. [Collected Works Vol. 3 (2021) Part I Chapter 3]
Yamazaki R. (2014b) *Agricultural structural dynamics under globalization*, Ochanomizu Shobo,
Yamazaki R. (2019) Controversy over the concept of primitive accumulation of capital, *Journal of Rural Issues*, 84. [Collected Works Vol. 5 (2022) Part I Chapter 3]
Yanagida K. (1942) *Japanese festivals*, Kadokawa Bunko.
Yanagida K. (1961) *Maritime Road*, Iwanami Shoten.
Yasuda K. (1971) *Russian Revolution and the Mir Community*, Ochanomizu Shobo.
Yorozuya S. (2016) Historical and theoretical development of the “problem in the economic field” (Part 1), *Sapporo University Journal of Economics and Management*, 46 (1, 2), pp. 107-159.
Yoshida M. (ed.) (1975) *Agriculture and land tenure in Africa*, Institute of Developing Economies.
Yoshida M. (1999) Actor in East Africa’s rural transformation and land system reform: Focusing on Tanzania, Ikeno S. (ed.) *Reexamining the image of African rural areas*, Institute of Developing Economies, pp. 3-58.
Yoshioka A. (1981) *Modern British economic history*, Iwanami Syoten.

Index

Ryoichi Yamazaki

(1957-)

Professor Emeritus at Tokyo University of Agriculture and Technology

Agricultural Economist

Between 2020 and 2022, complete five volumes of Ryoichi Yamazaki's Collected Works were published by Tsukuba Shobo.

Since 2022, he has been running the YouTube channel "Agricultural Economics and Beekeeping."

Primitive Accumulation and Community
by Ryoichi Yamazaki

2024年12月25日　第1版第1刷発行

著　者　山崎 亮一
発行者　鶴見 治彦
発行所　筑波書房
東京都新宿区神楽坂2－16－5
〒162－0825
電話03（3267）8599
郵便振替00150－3－39715
http://www.tsukuba-shobo.co.jp

定価はカバーに示してあります

印刷／製本　平河工業社

ISBN978-4-8119-0688-1 C3061